IEE MANAGEMENT OF TECHNOLOGY SERIES 20

Series Editors: J. Lorriman
G. A. Montgomerie

Skills development for engineers

Other volumes in this series:

Volume 1 **Technologies and markets** J. J. Verschuur
Volume 2 **The business of electronic product development** F. Monds
Volume 3 **The marketing of technology** C. G. Ryan
Volume 4 **Marketing for engineers** J. S. Bayliss
Volume 5 **Microcomputers and marketing decisions** L. A. Williams
Volume 6 **Management for engineers** D. L. Johnson
Volume 7 **Perspectives on project management** R. N. G. Burbridge
Volume 8 **Business for engineers** B. Twiss
Volume 9 **Exploitable UK research for manufacturing industry** G. Bryan (Editor)
Volume 10 **Design and wealth creation** K. K. Schwarz
Volume 11 **Integration and management of technology for manufacturing** E. H. Robson, H. M. Ryan and D. Wilcock (Editors)
Volume 12 **Winning through retreat** E. W. S. Ashton
Volume 13 **Interconnected manufacturing systems** H. Nicholson
Volume 14 **Software engineering for technical managers** R. J. Mitchell (Editor)
Volume 15 **Forecasting for technologists and engineers** B. C. Twiss
Volume 16 **Intellectual property for engineers** V. Irish
Volume 17 **How to communicate in business** D. J. Silk
Volume 18 **Designing businesses: how to develop and lead a high technology company** G. Young
Volume 19 **Continuing professional development: a practical approach** J. Lorriman

Skills development for engineers

An innovative model for advanced learning in the workplace

Kevin Hoag

The Institution of Electrical Engineers

Published by: The Institution of Electrical Engineers, London,
United Kingdom

The Institution of Electrical Engineers,
Michael Faraday House,
Six Hills Way, Stevenage,
Herts. SG1 2AY, United Kingdom

British Library Cataloguing in Publication Data

Skill development for engineers:
an innovative model for advanced learning in the workplace. –
(IEE management of technology series; no. 20)
1. Engineers – Education (Continuing education)
I. Hoag, Kevin L. II. Institution of Electrical Engineers
620'.00715

ISBN 0 85296 979 1

Typeset by RefineCatch Limited, Bungay
Printed in England by The Cromwell Press, Trowbridge

Contents

Acknowledgements

At the time work on this book began I was responsible for managing the continuing education programs for the technical staff of Cummins, Inc., in Columbus, Indiana, USA. My colleagues and I at Cummins had the good fortune to have our work honored with the first *Glenn L. Martin Award for Corporate Leadership in Continuing Engineering Education*, presented by the International Association for Continuing Engineering Education (IACEE).

Shortly after the award was presented at the IACEE conference I was approached by John Lorriman of Knowledge Associates, and asked if I would consider writing a book. My acknowledgements must begin with a hearty thanks to John for the vision for this book, and his helpful role as Series Editor for the Institution of Electrical Engineers throughout the project. As the manuscript proceeded several others at the IEE provided guidance, and were faced with the task of converting my scrawling into a presentable text. My thanks go out to Robin Mellors-Bourne, Publisher, Roland Harwood, Commissioning Editor, and Diana Levy, Production Editor.

The support and encouragement of Cummins Inc. was also very much appreciated. I especially wish to thank Jerry York, my colleague during the time that many of the ideas described in this book were developed. His work and ideas are greatly reflected throughout the book, and his review of the manuscript was invaluable. Others at Cummins to whom I am indebted for their encouragement, input, and permission to use the various examples include Jean Blackwell, John Wall, Roy Primus, and David Cole.

I owe many thanks to Professor Alan Knox of the University of Wisconsin. His guidance in the field of adult education, and his detailed review of my manuscript were greatly appreciated.

Finally, I must thank my family, Christine, Erin, Bethany, and Amy. The time required to complete this project seemed to continually grow beyond expectations, and their understanding and encouragement allowed the work to see completion.

Kevin L. Hoag
Madison, Wisconsin
April 2001

List of figures

List of tables

PART 1

AN INNOVATIVE MODEL

When asked about continuing education in the workplace almost anyone immediately thinks of classroom learning. However, classroom learning is best suited to convey relatively basic information to large groups of people – what is often most needed in continuing education is for relatively small numbers of people to develop advanced proficiencies in very specialized subjects. The first part of this book is intended to guide a shift in thinking toward methods that support advanced, specialized learning.

The topic begins with the need to reorient the thinking of an organization's leadership – away from the classroom, and toward alternative approaches. The primary emphasis of Part 1 of this book (Chapters 1 through 7) will then be on developing and implementing a model that guides and supports advanced learning. The roles of upper management, the direct supervisor, and the individual will each be covered. Techniques for creating the model, and communicating it throughout the organization will be presented.

Part 2 (Chapters 8 through 10) will then address further applications that follow directly from this alternative model. The book concludes with brief concluding remarks in Chapter 11.

Chapter 1

Introduction

Any Company, Inc. (a fictitious, but typical corporation) is an international manufacturer of mechanical and electronic devices. While this company has been in business for over 100 years, increased worldwide competition has recently squeezed its profit margin, and significantly decreased its ability to differentiate its products from others on the market. As would be expected of a company with a reputation for innovative leadership, Any Company has responded to its changing environment with new work systems and a revised organizational structure. These moves were designed to improve product quality, better understand customer needs, and reduce product development time and costs.

It was quickly recognized that implementing these new work systems would significantly change the day-to-day work expected of most Any Company employees around the world. The new systems would increase the need for employees to work together in a team environment, and for each employee to take on greater decision-making responsibility. Every employee would be required to learn the concepts and application of total quality system procedures. Every employee would be expected to gain a greater understanding of how Any Company products are used by its customers, and what the customers expect from Any Company. Finally, it was recognized by the astute Any Company leadership that technological change was occurring at an ever more rapid pace, and that remaining competitive would require employees to continually learn the application of new technology. In short, Any Company was faced with the need to create a learning environment, and ensure that employees gained new knowledge, skills and abilities throughout their careers.

The senior management at Any Company was quick to respond to the needs just summarized, and they assigned a small team to develop and implement a strategy to provide worldwide employee training where and when needed. Over a period of several years this team assembled an impressive array of training programs, supporting infrastructure, and part- or full-time training leaders at each of Any Company's worldwide locations. Standardized formats and approaches were developed; course masters were maintained, and a distribution network created; instructor certification programs were developed; an employee training tracking system was implemented; and an impressive catalogue of several hundred courses was created. From all appearances Any Company was quite successful in creating its corporate university.

Things were going along quite nicely until Any Company began to see evidence of an economic downturn on its horizon. To ensure that the necessary measures were taken early, operating expenses were scrutinized a bit more carefully. Several anecdotal items suggested to senior management that training costs might be getting out of hand. A consulting firm had been contracted to develop a series of interpersonal skills courses; an outside publishing house had been contracted to print course materials; floor space had been leased for a training center. These line items led to further scrutiny and a study team was assigned to assess Any Company's true training costs. While this proved to be an impossible task, several further findings emerged. First, the training function had grown significantly from the small initial team. Several plants now had training organizations and there was evidence of duplication of effort. Second, no one really had the time to develop all the needed courses, so in order to make rapid progress, a significant portion of the course development and instruction was being contracted to outside firms. Third, there were virtually no measures of the cost effectiveness of any of the training programs. A few studies had been done which demonstrated that employees who participated in particular courses learned some new things. However, it had been very difficult to determine whether this resulted in improved work performance, much less whether the amount of improvement justified the costs of the course and lost productivity while the employee was in the classroom.

While the study was not successful in identifying the total costs of training, the evidence was sufficient to conclude that training costs would have to be contained. The Any Company leadership, wishing to act decisively with this new knowledge, immediately re-deployed over half of the full-time training staff and halted most outside contracting. A clear message was given to the training leaders to reduce costs wherever possible, scale back the training offerings, and implement training effectiveness measures.

Any Company is a fictitious company, but the sequence of events described in this story will be all too familiar to many professionals involved in corporate training roles:

- the competitive environment which drives a need for continuous employee skill development;
- the creation of a corporate training center or 'university';

- the questions raised regarding training effectiveness;
- the start-and-stop approaches to corporate training.

All of these events are repeated time and again at corporations in a variety of industries around the world.

This book begins by questioning the approach taken by Any Company and so many organizations to address their employee needs for continuous learning. Is the development of a corporate training center, or 'corporate university,' the best method of ensuring continuous learning? Or are there alternative models that may be less costly and more effective? The purpose of this book is to present one such alternative model.

The learning model to be presented here builds on the concepts of experiential learning – employee-led learning occurring on-the-job, complemented by a variety of learning interventions particular to an individual's need. Several specific elements are made available to employees in order to instill the expectations of continuous learning and provide opportunities for self-directed, on-the-job learning. These elements include detailed guidelines and recommendations developed by leaders within the organization; a **hypertext** database making this information readily available to all employees; and planned career development programs for various employee groups.

The model that will be presented is applicable over a wide range of private- and public-sector organizations. Specific examples will be presented from the author's experience of its application at Cummins, Inc., an international producer of diesel and gas engines and power systems. The application of this model resulted in Cummins receiving the first triennial Glenn L. Martin Award for Corporate Leadership in Continuing Engineering Education, from the International Association for Continuing Engineering Education. The model has also been demonstrated to meet ISO 9000 certification requirements for employee training.

In setting the stage for the learning model that will be presented, Chapter 2 provides a serious discussion of the needs faced by industry, and the types of learning interventions that can best meet those needs. At the core of the model is the commitment of senior management to provide useful direction to their employees. Chapter 3 discusses the roles of management, and the specific contributions they must make to the success of employee learning. Chapter 4 provides further detail on recommended mechanisms to support learning in the workplace – specific decisions and selections that must be provided by the leadership within any given discipline.

With Chapters 2 to 4 as background, the information framework of the model is presented in Chapter 5. The model requires an effective communication mechanism to make up-to-date information available to all employees. This chapter will include discussion of **hypertext** technology, and various tools and approaches that can be taken to structure the required database.

NOTE: If you are a reader who normally skips the introduction, please make an exception and read this part. It is especially important to the remainder of the book!

Before proceeding further it will be necessary to clearly define several terms that will be used throughout this book. They are not terms new to the English language (or even to its North American form as used by this author!). They are, however, terms widely used in the field of adult education and have specific meanings within that field, and they are terms that will be quite central to the topics addressed in this book.

The core topic of this work is learning – specifically, adult learning within the industrial workplace. An important contribution to the terminology associated with learning is the now classic study edited by Bloom (1956), *Taxonomy of Educational Objectives.* In that work he identified three domains of learning – Cognitive, Affective, and Psychomotor. These domains are now most often referred to by their descriptors, Knowledge (cognitive), Skills (psychomotor), and Attitude (affective), or KSA. When adult learning in the workplace is discussed, the reference is to some combination of knowledge, skills, and attitude, with the majority of the emphasis placed on either of the more measurable or quantifiable domains, knowledge or skills. In this book the term, **learning** will be used to represent growth in various combinations of these domains.

As the model presented in this book is developed it will often be important to talk about the various topics or areas of study in which individuals are expected to possess growing capabilities as they proceed in their careers. Each of these topics will be referred to as a **subject** of learning.

It will be important to be able to refer to the level of capability an individual possesses in a particular subject. Measures of capability in a given subject will be termed **proficiency**.

Finally, central to the learning model presented in this book will be the development of matrices presenting the combination of subjects and the proficiencies needed in those subjects for any particular position within an industrial organization. A **learning matrix** will be the term used to describe a matrix of the subjects and proficiencies expected of an individual holding a particular role.

Fundamental to this learning model is self-direction. Each employee holds primary responsibility for their own needs assessment and continuous learning. Chapter 6 details this expectation, and discusses specific aspects of the model that support individual responsibility. While the individual holds the primary responsibility for their learning, this does not let management off the hook, and Chapter 7 emphasizes the roles that direct supervisors must play in supporting their employees' learning.

A fundamental element of continuous learning is experience. While one can gain the depth required for specialized roles by gaining project experience within a particular discipline, rotational programs may be required to provide the experience across disciplines that will be required for project leadership roles. Chapter 8 discusses the use of rotational programs – not as orientation tools, but in a managed career progression.

The next two chapters identify further benefits that follow quite naturally from the learning model. Chapter 9 discusses the model's use in organizational assessment and resource planning. Chapter 10 articulates its direct use in evaluating the effectiveness of particular training programs or other learning interventions.

The book concludes with some final considerations pertaining to the application of this model. Several tips gained from the author's experience in developing and fine-tuning the model are presented.

Chapter 2

Moving beyond the classroom

The importance of lifelong learning cannot be seriously questioned. Not only within industrial careers but in all aspects of life, the rapid pace of technological change requires everyone to continue learning throughout life. As portrayed in Chapter 1, manufacturers and other businesses are faced not only with technological advance, but also with competitive needs to significantly change their organizational structures and the ways in which they work. Accordingly this discussion begins by taking the need for continuous employee learning as a given.

This chapter begins by looking carefully at several criteria important to the ongoing learning process in industry. Attention then turns to the various popular training methods and tools, paying special attention to their ability to address the criteria that have been identified. Several specific examples will then be provided to demonstrate how these various criteria impact the selection of training methods and tools. Two conclusions will emerge from this example. The first should not be surprising to most readers – careful attention to the objectives of a given learning intervention goes a long way in selecting the most effective training methods and tools. The second, and more important conclusion for the emphasis of this book, is that **none of the typical approaches to training in the workplace are at all well suited to some of the most important professional development needs**. This chapter will conclude with a detailed look at the progression of learning – how a person moves from rudimentary knowledge of the basic terms and applications of a given topic, to where she or he is able to teach others its detailed application, and ultimately becomes an acknowledged expert in a particular field. This discussion is a necessary starting point to begin thinking about learning programs that will effectively support the higher levels of this learning progression.

2.1 Taking a step back

When most people think about the learning process the first picture that comes to mind is that of a school building or a classroom. After all, our formal education is certainly a universal and important part of our learning experience. It is perhaps only natural therefore that when corporate managers recognize a need to ensure continuous employee learning they too immediately think of classrooms, learning centers, and ultimately 'corporate universities.' In some cases this may well be the appropriate response. However, this is not the only approach that can be successfully implemented to address employee learning needs. Depending on the specific needs of an organization, the creation of a corporate training center may in fact be exactly the wrong approach. Before proceeding further into the development of a training center it is important to take a step back – to take a broader look at the learning needs of the organization, and the best ways to meet those needs.

It will be helpful to begin by identifying some of the considerations that will be of importance to the organization. The following criteria represent a general set of considerations that will be applicable to virtually any organization.

2.1.1 Audience size

Perhaps the first question to be considered is that of audience size. Most training programs will become more cost-effective (increased return on investment) as audience size increases. Large organizations are often able to justify training programs or training centers that could not be considered by smaller firms. This point should be quite obvious, however a second aspect of audience size is often overlooked. Even within large organizations many learning needs are faced by a very small fraction of the employees. A large company may have a very well regarded corporate university, but may be missing the mark on the continued professional development in subjects critical to the organization – or may be taking a terribly cost-ineffective approach to addressing those learning needs.

2.1.2 Audience location

Clearly another important consideration is that of audience location. Large corporations often face the need to rapidly and cost-effectively provide for development needs of employees located all over the world. Classroom training becomes very costly when instructors or students must travel great distances to attend needed training programs. Rapid growth in the development of various distance learning technologies attests to the need for cost-effective skill development programs at geographically dispersed sites.

2.1.3 Required proficiency

Another very important employee learning consideration is that of the required proficiency of the employee in the particular subject. The first steps in identifying the specific objectives for employee learning are to decide both what needs to be learned and how well it needs to be learned. Table 2.1 summarizes the general progression of learning a person experiences as a new subject is approached. It is a progression in which we all participate every day of our lives. However, it is a progression that is seldom discussed by corporate training leaders today, and it is all too often ignored in the creation of training programs. This topic will be taken up in greater detail later in the chapter.

2.1.4 Technical difficulty

The technical difficulty of the subject matter is often an important consideration in selecting one's approach to employee development. This consideration plays out not only in identifying individuals within the organization, or outside contractors, to develop the course, but in ensuring that instructors will be available to teach the material throughout the life of the course. This becomes especially difficult when dealing with geographically dispersed employees needing to develop proficiency in the subject. Can the necessarily expert instructor be counted upon to always be available to teach the course where and when needed?

2.1.5 Ease of change

As one begins to review the various commonly used methods and tools of training it will be important to consider how often the material will need to be revised. Is the subject one for which little change occurs over time? Or is it one that, due to the rapid pace of technological change, or rapid changes occurring with a product, a process, or within the organization, will require frequent updates to the training material? This question has serious implications for the selection of training methods.

2.1.6 Cost

For most organizations today this particular criterion should probably be placed at the beginning rather than the end of the list. Clearly the cost-effectiveness of any learning program will be increasingly scrutinized as operating environments become more and more competitive. Some costs are easily tracked, while with many programs there are hidden costs that may be draining company resources, but are not being tracked. For example, an outside contractor might be hired to develop a highly technical course on a subject in which that contractor is a

Table 2.1 Classifications describing the learning continuum

Proposed learning continuum	Bloom's taxonomy	Knox
Terminology, definitions Context and fundamental rules	1.0 Knowledge 2.0 Comprehension	*Fragmentary understanding.* Learner processes facts and details superficially, without central concepts or broad integrating themes, and considers or describes only one side of a problem or issue with little attention to similarities, differences and gradations.
Integration with prior knowledge Basic application and refinement of learning Proficient application in commonly seen situations	3.0 Application	*Comprehension.* Learner deals with central concepts or integrating themes to recognize similarities but without relation to supporting facts and details that help distinguish differences. Opposing views are perceived as compartmentalized or negative.
Further depth in fundamental understanding, and learning limitations Ability to apply over increasingly broad range of applications	4.0 Analysis	*Understanding of relationships.* Learner integrates concepts and themes with facts and details, by identification of both similarities and differences and deep multiple processing, within the context of information presented.
Learning special considerations for difficult applications Increasing involvement in teaching others Contribution to expert dialogue Participation in cutting-edge research Leading independent research	5.0 Synthesis 6.0 Evaluation	*Inclusive Understanding.* Learner uses deep processing of integrating themes to go beyond the context of information presented, using personal experience and additional knowledge to provide reasons for similarities and differences and to explore relations among alternative views, which contributes to better retrieval and new insights.

recognized expert. However, the contractor is not very familiar with the specific application of the subject in this particular industry, and asks for a contact within the company who might be able to help with 'a few questions.' An excellent course is developed, within the contracted budget. However, unbeknownst to the training staff those 'few questions' resulted in hours of telephone conversations, calculations and graphs, and several face-to-face meetings. The true costs of developing the course may easily have been two or three times those that were tracked.

A still more difficult aspect of calculating the return on investment is that of measuring the effectiveness of the resulting program. How many people learned how much? How did this new learning help them in their work? What were the cost savings resulting from the increased employee capabilities? These are difficult measurements to make, and they are seldom if ever done well.

2.2 Approaches to provide for learning in the workplace

Several of the most common approaches to training will now be briefly discussed, with specific emphasis on their advantages and disadvantages relative to the criteria identified in the preceding section. Table 2.2 provides a tabular overview of the discussion that will then be expanded in the sections that follow. In the table, each one of the criteria from the previous section is listed across the top. The various approaches to training are listed in the left column. The ability of each approach to address the various criteria is briefly summarized in the other columns. It should immediately be noted that all of the approaches listed here are most cost-effective when developed for large audiences. It should also be noted that they are all best suited to address fairly low proficiency levels. We will return to these two criteria in greater detail later in this chapter. Some differentiation between the approaches is seen as they are evaluated versus the remaining criteria.

2.2.1 Classroom training

The most traditional, and still most often utilized approach to training is the formal classroom. Interestingly, while there is today a great push to move away from classroom training to less traditional methods, almost the entire argument is based on making training available where and when it is needed – in other words, addressing the problem of audience location. Little is said about the relatively low proficiency levels it addresses, probably because most of the non-traditional approaches share this disadvantage. With technically difficult material, it is important to ensure that an expert instructor is secured every time and in every location the course is taught. Instructional material should be easily

Table 2.2 Summary of the ability of various training tools to meet important employee development criteria

	Audience size	Audience location	Required proficiency	Technical difficulty of material	Ease of update	Cost to develop and use
Classroom	Larger audience more cost effective	Poor for dispersed audience	Low	Requires close attention to instructor selection	Good	High usage costs
Teleconference	Larger audience more cost effective	Expensive for dispersed audience	Low	Good	Good	Very high usage costs
Video tape	Larger audience more cost effective	Good for dispersed audience	Low	Good	Poor	Low usage costs
CD-ROM	Larger audience more cost effective	Good for dispersed audience	Low	Good	Poor	High cost to create; low usage cost
Intranet	Larger audience more cost effective	Good for dispersed audience	Low	Good	Good	Moderate cost to create; low usage cost

updated. While course development costs are generally similar to those associated with most other approaches, some costs are repeated each time the course is offered.

2.2.2 Teleconference

Teleconference systems in which both audio and video signals are transmitted in real time over telephone lines, while remaining expensive, are becoming popular in the larger organizations. Software is now available to transmit these same signals into personal computers, and the technology is thus gaining attention as a training tool. It holds the advantage of real time interaction with the instructor and other students, but the required hardware and software makes it one of the more expensive approaches, and it is expected to fall out of favor with the growth of CD-ROM and the internet.

2.2.3 Videotape

One of the earlier distance learning technologies, videotape remains at the time of this writing the most popular. It is relatively inexpensive, provides excellent audio and video quality, and allows each individual the opportunity to view the material where and when desired, albeit by forfeiting interaction with the instructor.

2.2.4 CD-ROM

The CD-ROM is actually quite similar to videotape in many of its advantages and disadvantages. Its greatest advantage over videotape is the ability to make the material more interactive, and allow the student to select various learning paths. It shares the videotape disadvantage of needing to be re-recorded whenever the course material changes.

2.2.5 Intranet

The greatest recent change in distance learning technology has been the advent of the internet, which now threatens to displace each of the other technologies discussed here. **Intranet** refers to the use of internet technology within the firewall of an individual organization, thus providing their desired confidentiality. While the internet does not yet have the audio and video capability of videotape or the CD-ROM, it is projected by most experts to soon reach that plateau. It offers the same interactive capability as the CD-ROM, but material is generally found less expensive to develop, and changes can readily be made to the material, and instantly written over any previous material.

The preceding paragraphs have described the approaches commonly used by corporate training departments for the continued skill development of their employees. Most corporate training centers have been created focusing on classroom training, and encompassing the work summarized in Figure 2.1. At the heart of the work is the development and offering of specific training classes. The process relies on accurate identification of the employees' learning needs, and its usefulness is assessed by accurate effectiveness evaluation. Unfortunately in most organizations these initial and final steps are done much more poorly than the middle steps – if they are done at all. In many organizations the process is now being broadened to include the development of one or more of the distance learning technologies. While this type of training center may very well be effective for some organizations it holds several inherent disadvantages.

The first and most obvious disadvantage is cost. Creating a corporate training center literally involves creating a business within one's business – identifying part of the company's business as education and training, and in some cases operating as a profit center. The training center will require a full-time staff, classroom space, a registration and tracking

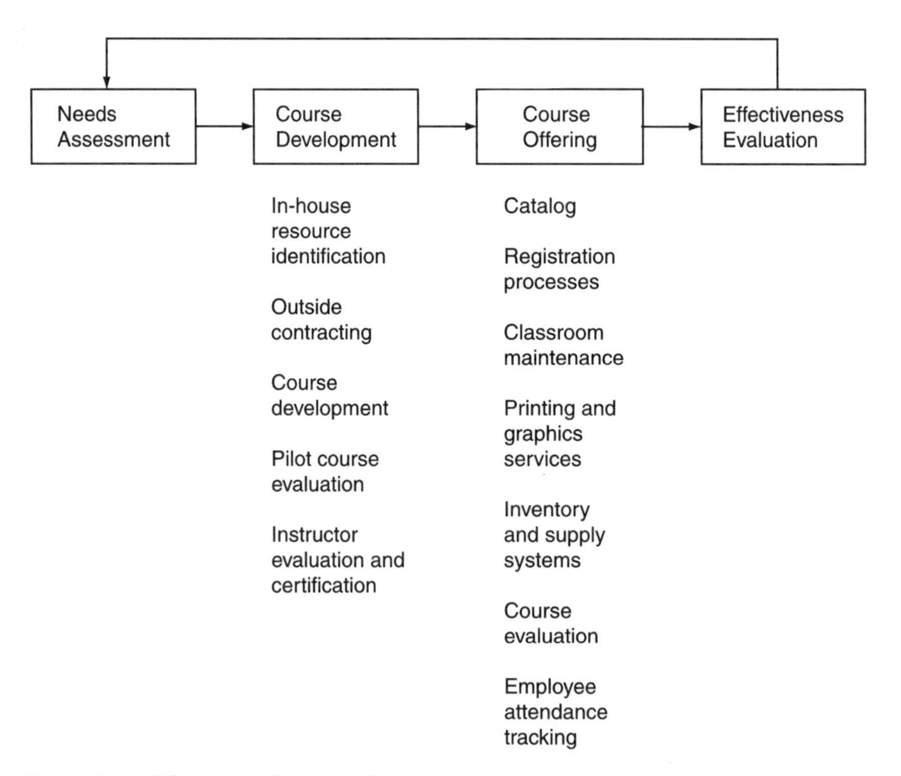

Figure 2.1 Elements of a typical corporate training center

system, and support for each of the functions listed in Figure 2.1 (and perhaps others as well).

The second, but seldom discussed disadvantage is the question of instructional expertise. Classroom and distance learning approaches to instruction are certainly important elements in the process of continuous learning. But all too often those who become the training leaders and instructors in corporate training centers possess little or no background in, or understanding of, adult education. They may learn a great deal as they go, but much of this learning happens at the expense of the class participants' experience, and may very well undermine the credibility of the training center before it gets off the ground. Even a cursory review of much of what is being taught in most corporate universities and consulting firms will suggest the need for much greater attention toward addressing this problem. This is not to say that universities and other specialists in adult education do not suffer from the same problem. But they are in the business of addressing that problem, and will need to do so to satisfy their customer base. Business and industry has its own set of customers and challenges . . . why divert resources to take on another set?

A third potential disadvantage of the corporate learning center, and one of great importance to the further discussion in this book, is the resulting proficiency focus. As discussed earlier with reference to Table 2.2, the proficiencies that can be expected from classroom or distance learning approaches are relatively low. The corporate learning center, focused on traditional classroom methods or on the modern distance learning technologies, spends a lot of money while not addressing the advanced proficiencies that might well be critical to the company's future.

2.3 Examples of employee development needs and how they might be addressed

The discussion now turns to some specific examples of employee development needs within organizations. A few situations that are very well served by classroom or distance learning approaches will be presented. Also presented will be some situations where these approaches will be found wanting.

2.3.1 New product familiarization

The company has just released a new product, and needs to rapidly provide information to a large number of sales and service personnel at various locations around the world. The initial requirement is for basic familiarity – an understanding of how this product differs from the one it replaces; an overview of its design features; perhaps a few notes on special considerations and tools required in servicing the product in the field. Historically such training was conducted in classrooms, often using a 'train the trainer' approach. A group of more senior people would receive the training first, and then take it to various locations around the world, 'cascading' the training to others. Training needs like this one are where the new distance learning technologies have made their most rapid inroads. A videotape, CD-ROM, or possibly intranet approach will be quite effective in meeting this need.

2.3.2 A new work system

The company has just developed a new work system, involving a reorganization, and changes in the processes and flow of work, from advanced product development to the shop floor. It will be important that everyone learns this new system rapidly. Classroom training remains by far the most popular approach for such communication – and certainly with good reason. The company needs to ensure that everyone receives the training, so they will probably register attendance. Employees will want to discuss how the changes will impact their own work, and may have questions for the instructor. The instructor(s), who may well be selected from senior management will want to use the classes to get a first-hand look at how the employees respond, and to note any potential problems that they will need to address.

2.3.3 Technical writing

Many employees in the company need to be proficient in written communication. Virtually all of these employees learned the basic terminology and rules of sentence structure in the early years of school (although we can certainly attest to the all too common need for remedial learning of this subject!). Most of the people are well along the path of applying their learning in their daily lives, having written letters to friends, and reports and essays in school. The goal of the organization in further developing employee communication skills is not to create expert linguists or communication professors – it is merely to expand their experience of proficient application to include business or technical writing. These proficiency needs must be clearly defined as the basis for deciding the organization's needs for skill development programs. What types of programs will best focus on increasing the employees' experience of written communication to include business or technical writing?

Almost every organization faces the need for such a program. Many companies have developed in-house programs, and several technical writing seminars are available through consulting firms and other contractors. However, almost none have seriously faced the question just posed – what type of program will best focus on increasing the employees' experience of written communication to include business or technical writing? The need is for guided experience, allowing employees to put their own experience of writing into practice. One approach might involve identifying several on-the-job mentors to help people with their technical writing needs as they occur. While it is always difficult to dedicate resources to such roles, it will be far less costly and far more effective than sending employees through the next two-day technical writing seminar.

2.3.4 Expertise on a specific subject

A critical aspect of the company's product durability is ensuring against fatigue failure under particular operating conditions. There is a small group of engineers dedicated to this problem, and they have developed some very effective techniques. However, one of their top people has recently retired, and another will retire in two years. At the same time, in an effort to reduce product development time and better meet local customer needs, more and more of the development work is occurring at plant sites around the world. There have recently been product failures attributed to the fact that these development groups have not applied the appropriate techniques.

The scenario just described will sound familiar to anyone working in engineering or management roles in a wide variety of industries. It is clearly an employee development problem, yet one that few (if any) corporate training centers are equipped to handle. At this point possible solutions to this example are not going to be discussed. The intent of the example is merely to suggest that many of the most critical skill development needs fall into this realm – a relatively small number of employees needing to gain an advanced proficiency on a specific subject. Most technology-based organizations face a great number of these needs at any given time, and the approaches to training discussed in the preceding section are quite ineffective in addressing the needs.

2.4 Developing technical expertise

The previous section concluded by identifying a critical employee development problem – the need to provide advanced learning opportunities on a large number of subjects, but with only a few individuals requiring any given subject. Thus the discussion returns to the training program criteria of audience size and required proficiency. In Figure 2.2 these criteria are viewed in terms of quadrants. Audience size is indicated by the vertical coordinate, and proficiency by the horizontal. As has previously been discussed, classroom learning and the commonly used distance learning technologies, are most cost-effective when developed for larger audiences. They are also best suited for teaching the most basic

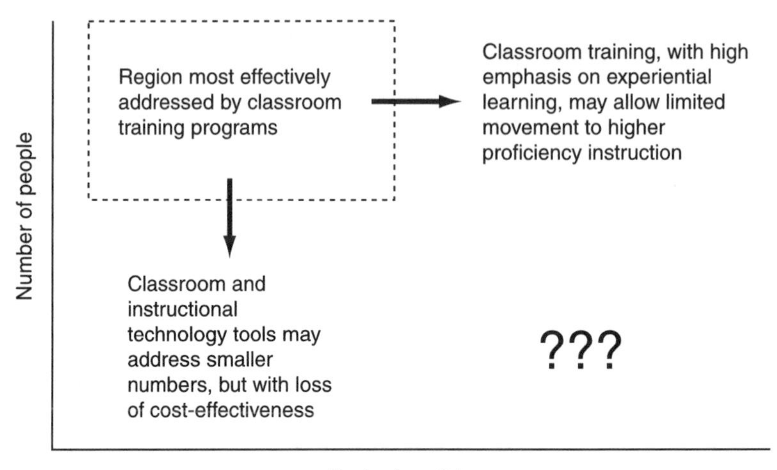

Figure 2.2 Training tool usage versus audience size and proficiency requirements

proficiencies of any given subject. The focus of these tools is thus in the upper left quadrant of Figure 2.2. As indicated in the figure, these tools can be applied for smaller audience sizes, but the cost-effectiveness of doing so rapidly becomes unacceptable. The development cost of a training seminar on any given subject is impacted only slightly, if at all, by the number of people for whom the material is developed.

Some training professionals would argue that the resulting proficiencies of the participants can be increased by incorporating experiential elements in our classroom seminars. I would agree, and insist that experiential elements should be incorporated wherever possible. However, I would also insist that while experiential elements work to increase participant proficiency they bring those participants nowhere near to the levels of proficiency often needed in technical organizations. This will become more apparent in the next section.

The question remains how to best address the skill development needs identified in the lower right quadrant of Figure 2.2. How can an organization ensure that the relatively small number of people needing advanced proficiency, in any of a wide number of subjects, are having their skill development needs met? This question should be of critical importance to the training and development staff supporting any organization.

In most organizations today the training staff is not involved in addressing this proactive question. A few organizations address the problem implicitly, through succession planning and staffing activities, but very few have explicitly identified the skills and proficiencies that are needed, or have addressed the means by which these skills might be developed in their workforce. The vast majority of organizations have no formal approaches, and rely on the experiences employees happen to gain throughout their careers, perhaps supplemented by a professional development seminar here and there. This is often supplemented by outside hiring to bring in the critical proficiencies that are found lacking in the organization.

It is striking to return to Figure 2.2 and note that the critical learning needs just discussed are in exactly the opposite quadrant from the focus of activities of most corporate training centers. The intent of this book is to present one model, and examples of its application, that shifts the focus of skill development from the upper left to the lower right quadrant of Figure 2.2. The discussion begins in the next section with a more detailed look at how people learn.

2.5 The learning progression

The processes by which we learn remain remarkably similar throughout our lives. On any new subject our growth in understanding, and the development of our abilities, progress through a continuum that can be represented by a relatively small number of classes of behavior. It is this observation that provides the basis for Bloom's taxonomy, as stated in the following quotation:

> It is assumed that essentially the same classes of behavior may be observed in the usual range of subject-matter content, at different levels of education (elementary, high-school, and college). Thus a single set of classifications should be applicable in all these instances.
>
> (Bloom, 1956)

The resulting taxonomy developed by Bloom and his associates consists of six major classes of learning, defined as follows:

- Knowledge
    - 1.0 Knowledge
- Intellectual Abilities and Skills
    - 2.0 Comprehension
    - 3.0 Application
    - 4.0 Analysis
    - 5.0 Synthesis
    - 6.0 Evaluation

Each of these classes is carefully defined in Bloom's original publication, and the reader is referred to this work for further study. For the purposes of the present study a learning progression applicable to professional development in the sciences and engineering is now presented. This progression has been summarized in Table 2.1. Also indicated in this figure are Bloom's six classes, and four progressively higher levels of understanding as identified by Knox in a study of adult learning (Knox 1986).

On any given subject there is a continuum of learning which occurs as people gain in their knowledge and capabilities pertaining to the topic. If it is assumed that the person initially has no knowledge of the topic whatsoever, learning begins with the person gaining exposure to the basic terminology and definitions pertaining to the topic. This is followed by the rudiments of the subject's importance and application. Especially with technical or scientific subjects some understanding of fundamental laws, governing equations, and limitations of its application will need to follow.

As the learning process continues the person will begin applying her learning, synthesizing this learning with what has been learned before, and testing the concepts through practical use. Initially the learning will be applied on its most basic levels. The person may have many questions and a need to turn to others for help. She may find that something has been misunderstood, or in some way needs to be corrected or adjusted when applied. Through repeated use, and application of the learning over a variety of situations, the learner will become increasingly proficient in the day-to-day application of the learning. This person could then be depended upon to apply the learning independently over an increasing variety of situations.

Continued application of the learning will result in the person coming across more sophisticated or difficult situations, where she will again have to turn to expert help. Each such application contributes further to the learning, and throughout this process the person may be expected to gain further knowledge of the theoretical underpinnings of the subject, as well as its practical application. Not only does the person become increasingly proficient in applying the subject, but she now possesses the expertise to teach others – either in a formal classroom environment, or through working with others in their efforts to apply their own learning.

Depending on the person's continued interest in the subject matter, and her needs in its application, she may begin pushing the existing boundaries of the subject – experimenting with its application in areas not previously considered, and perhaps expanding the fundamental knowledge base pertaining to the subject. This phase of learning involves dialogue with colleagues often from around the world. New information at the cutting edge of the subject is shared, challenged, proven and published. The person is now recognized by her colleagues as an expert in the field.

The learning process described in the preceding paragraphs is intended to portray a general progression that applies to everything from basic interpersonal skills to the most advanced topics in science and engineering. The first task of the training leaders in an organization is to identify the subjects on which each employee in the organization will need to gain knowledge and skill. The second task is to decide the proficiency within the learning continuum needed by the employee on each particular subject in order to effectively contribute to the organization.

Looking at the realm of engineering and scientific subjects, a healthy organization requires that the majority of practitioners are well into the practical application phase in their learning on subjects regularly used by the organization. A core of the more senior practitioners will have been exposed to a wide variety of the more difficult applications of the subject, and thus able to provide guidance and instruction to others. Leading organizations in most industries should also expect to have members recognized as world-class experts in most or all of the subjects most fundamental to their work. These people can provide the continued guidance and technical leadership that will facilitate the further development of the entire workforce, and create the catalyst from which new products and technologies are developed. Clearly most of the learning that must occur in the organization will be in the efficient application of the subjects to current work needs, and at the more advanced levels of application of the subject to new situations.

Figure 2.3 repeats the portrayal of the generalized learning continuum as originally shown in Table 2.1. Listed to the right of the continuum in Figure 2.3 are some of the commonly used tools through which learning takes place. For each subject required of each employee in an organization this continuum, and the most effective tools for learning at the desired proficiency levels, must be carefully considered. While organizations often turn first to classroom training it may well be that the greatest learning needs are for levels of proficiency that cannot possibly be addressed in the classroom. Training leaders and corporate managers often see their work as developing and coordinating classroom programs, while ignoring these higher proficiency needs. For many organizations it would be far more effective to focus limited training resources on learning programs to address these higher proficiency needs. While this need should be especially clear for engineering centers and scientific laboratories the same conclusion may hold true on the shop floor or among the sales and service staffs as well. It is critical to the organization to carefully assess where its resources in support of learning can best be applied. Figure 2.3 suggests that the higher proficiencies of greatest value to an organization are best developed through personal application, one-on-one mentoring, and dialogue with experts. The remainder of this book focuses on learning models that create the climate and infrastructure necessary to support the development of these needed proficiencies.

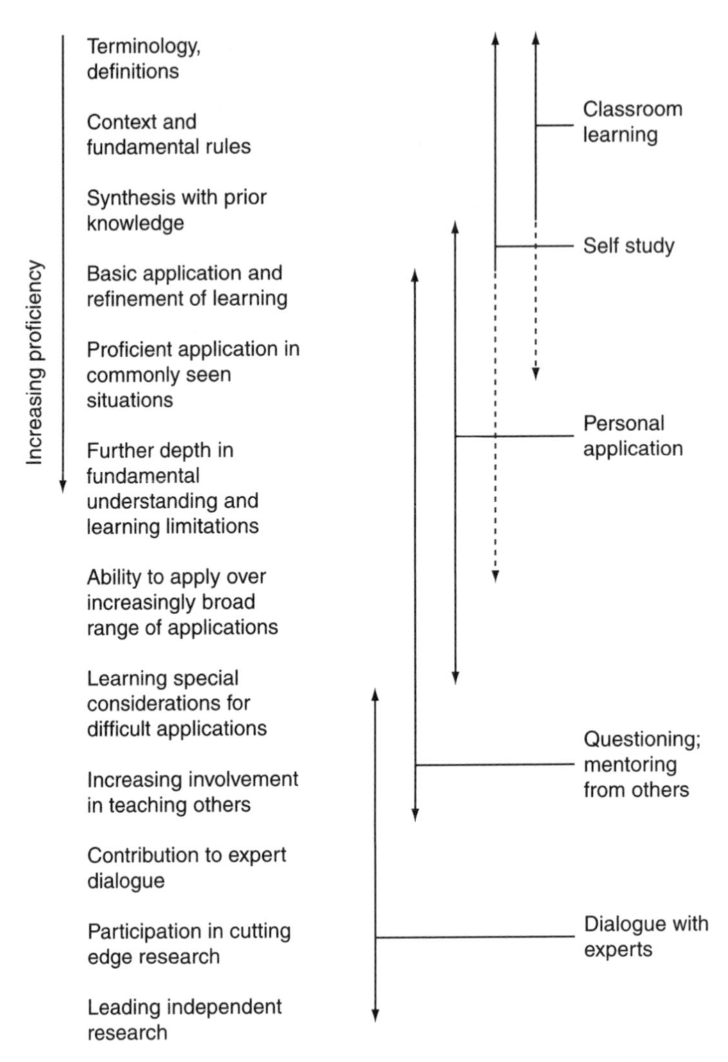

Figure 2.3 The learning continuum, with approximate realms of use of various skill development mechanisms

2.6 Proficiency scales

While the learning process is a continuum, the practical needs of society or any given organization will require some means of measuring and quantifying proficiency. A scale is necessary in order to identify current individual abilities and development needs. We are all familiar with the testing and grade assignment systems used in formal education. While this is a familiar example of efforts to measure proficiency it is typically not an appropriate method for a business or industrial organization.

An alphabetic or numeric scale, dividing the learning continuum into a series of discrete proficiencies, will be needed. The number of proficiency levels will be decided by the needs of the organization. While as many as ten distinct levels are often used on such scales, three or four levels will be very effective in most cases, with the emphasis placed on making distinctions at the proficiency levels of greatest importance to the organization. For example, most organizations will find it important to be able to quickly and accurately identify those employees having the abilities to independently lead projects requiring proficiency in particular subjects, or to teach others those subjects. They will have far less need to distinguish among practitioners at various early levels of application of the subject.

Later chapters will turn in more detail to the ways by which the proficiencies of a person are accurately assessed. The model that will be presented will emphasize creating an accurate means of self-assessment.

2.7 Summary

- The following criteria were identified as important considerations in selecting appropriate training methods:
  - audience size;
  - audience location;
  - desired resulting proficiency;
  - technical difficulty of the concepts being taught;
  - ease of change of the material;
  - cost.
- Much attention in industry today focuses on using technology to provide alternatives to classroom learning:
  - teleconference;
  - videotape;
  - CD-ROM;
  - intranet.
- Each of these tools is most appropriate for large audience sizes, geographically dispersed participants and relatively low desired proficiencies.
- Of critical importance to engineering and scientific organizations is developing and maintaining expertise in rapidly changing key technologies.
- The most important skill development needs of the technical organization are 180 degrees out of phase with the current emphases in corporate training. The need is for approaches best suited to educating small

numbers of people in each of a wide variety of key technologies, to very high proficiencies.

- The ways in which people learn are remarkably similar across subject matter and over the course of our lives. As proficiency increases, the learning occurs to a greater and greater extent through experience.

Chapter 3

Management roles

In the previous chapter it was argued that the most critical learning needs in an engineering organization often involve small numbers of people needing to gain very advanced proficiencies. A further summary of the engineering career development challenge follows:

- Many Topics – The products produced are often complex, and call on a variety of engineering disciplines in which the company must possess expertise.
- Small Number of People per Subject – A small cadre of experts in any given discipline ensures that others have a source for learning and growth, and that the company provides leadership in continued enhancement of that discipline.
- High Required Proficiencies – In order to accomplish the preceding, the company must ensure that mechanisms are in place to support the continuous development of expertise – keeping up with technical change and building future technical leadership.

The preceding chapter also pointed out that the need just described is not at all well served by the structure and priorities of most corporate training organizations. A quantum shift in the approach to learning in the workplace is required. Such a shift will necessarily involve the active support and leadership of the company's management. Thus the discussion must begin with management roles if changes in the approach to employee learning are to be addressed.

This chapter begins by addressing the specific problem of building management support. The roles of engineering management in creating the foundation of a new learning model are then described in successive sections. Their active involvement in identifying the types of positions, identifying the subjects in which learning is required for those positions, and finally, identifying a required progression of proficiency will each be taken up in turn. Each section will begin with a general discussion of the principles and objectives. This discussion will then be followed by several specific examples.

3.1 Defining the requirements

In order to support advanced learning the emphasis must be placed not on traditional classroom activities but on creating the mechanisms to encourage and support learning through on-the-job experience, one-on-one mentoring, and the discussion and application of new techniques in the workplace. These learning mechanisms should be supplemented with self-study and in some cases, classroom seminars – especially those with a significant hands-on component. Clearly the engineering management in any given organization will play a key role in creating and sustaining the needed environment for such advanced learning to take place. It will be up to management to create the definitions of expertise required in each engineering role, and to set the expectations of continuous learning for each person, and of the support required by each engineering supervisor. The following headings summarize the roles that will be required of each participant.

3.1.1 Senior engineering management

- Set expectations regarding the subjects and proficiencies required in each engineering role.
- Provide non-threatening, developmental tracking mechanisms.
- Clarify the expectations placed on the engineering workforce and their direct supervisors, as defined below.
- Communicate recommendations and guidance for developing proficiency in particular subjects.

3.1.2. Individual engineers

- Self-assessment of capabilities and development needs versus the expectations defined by management.
- Utilize the approaches recommended by management, and others identified by oneself, colleagues and supervisors, to take responsibility for personal learning.

3.1.3 Direct supervisors

- Provide project assignments that will build on each engineer's experience while allowing them to gain proficiency in subjects consistent with their individual career interests and the needs of the organization as a whole.

- Allot time and resources required to support the new learning. The primary learning mechanisms of this model are accomplished on the job, but this may need to be supplemented with advanced seminars, for example, that must be made available as needed.

3.2 Establishing management support

Having presented the objectives of advanced learning, and having taken a first step toward clarifying the specific roles of senior management, the individual engineers, and their supervisors, we are now faced with the task of establishing the management support that will be required to implement this learning model. The purpose of this section is to provide the key elements that must be communicated to management in order to gain their support. Two elements must be communicated. The first is to briefly summarize the importance of an alternative model – in other words create one or two slides that summarize the main points made in the second chapter of this book. The second is to succinctly present the alternative model. In the spirit of a brief tutorial, the remainder of this section presents a proposed slide presentation that might be useful in presenting these concepts to management for their discussion and support. The proposed slides are shown in Figures 3.1 to 3.5. Of course the specific content of the slides may be modified to address the needs of a particular organization.

At the risk of leading the reader to wonder why it took so many words to get to this point, Figures 3.1 and 3.2 are intended to summarize why an alternative learning model is important. Figure 3.1 summarizes the make-up of an ideal engineering organization. This ideal is widely

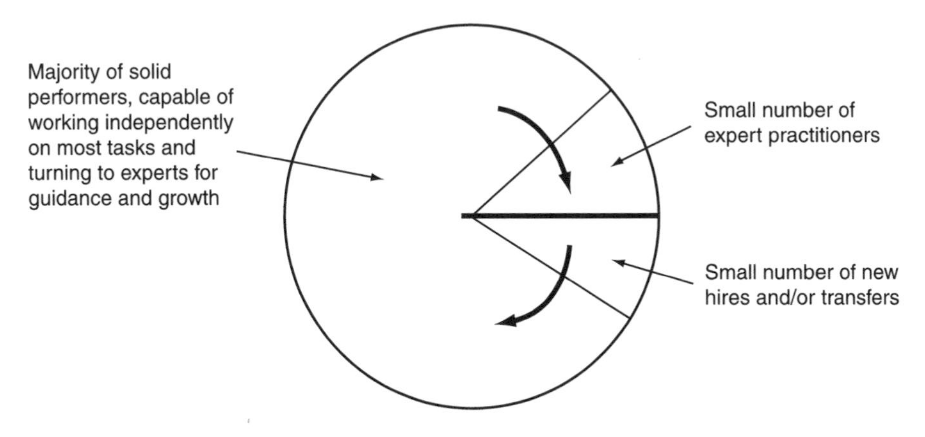

Figure 3.1 The make-up of an ideal engineering organization

- Bring new hires and transfers quickly up to speed
  - Integration into body of solid performers – *Done well by most organizations*
- Continue development of the majority of engineers, to ensure succession to expert proficiency takes place

Figure 3.2 Employee development focus

accepted, and should generate little or no argument. The statement being made is that a healthy engineering organization, at any point in time, should include a few senior 'expert' practitioners, a large majority of solidly proficient engineers capable of independently completing most tasks, and a small percentage of new hires or transfers. In presenting this slide it should be emphasized that this desired make-up is applicable both to the corporation as a whole as well as to any individual engineering group or organization within the corporation. The discussion pertaining to this slide should conclude with reference to the arrows indicating the movement of people from one category to the next through their continued learning.

Figure 3.2 then provides an outline of the learning represented by the two arrows from Figure 3.1. The first point describes the integration of new engineers into the workforce. It is included as a point of reference – it could be stated that this is where most of the organization's formal training efforts are focused. The second point then emphasizes that another important aspect of employee development is to ensure that today's solid performers are becoming tomorrow's experts. It can be emphasized that this development effort is vital to the organization's long-term health, but that it is currently being left to its own devices, and is not being supported by the company's training efforts.

The next slide presented to management might be Figure 3.3, which is repeated from Chapter 2 of this book. The point to be emphasized is that the focus of the new learning model is to be on advanced proficiencies, and that within any given engineering discipline, the number of people needing to develop these advanced proficiencies is relatively small. The figure then points to the fact that traditional classroom methods are not well suited to the development needs of these people. The emphasis of the new model is to be in the quadrant indicated by the question marks, thus leading into the remaining slides, where a model intended to address this quadrant is presented.

Figures 3.4 and 3.5 complete the presentation with a brief summary of the proposed advanced learning model. These slides are sure to generate further questions and discussion. A cursory discussion of these slides is now given, and the remaining sections of this chapter provide a far more detailed look. These later sections provide the information required for

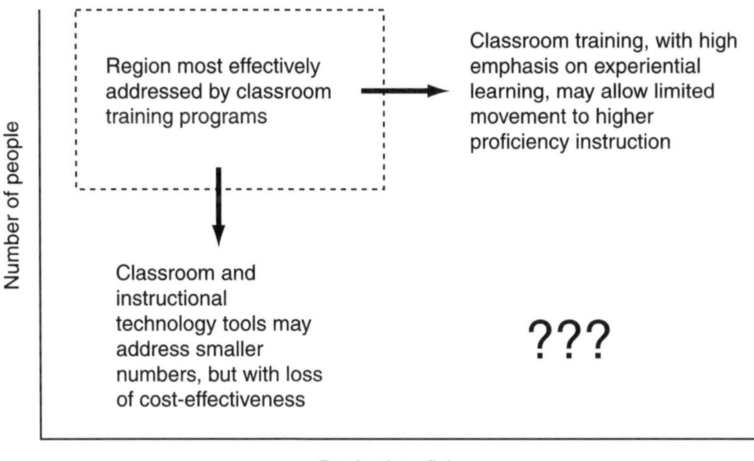

Figure 3.3 Training tool usage versus audience size and proficiency requirements

- Set clear expectations of subjects required and proficiency goals
  - Current needs
  - Future needs.
- Place responsibilities with each individual.
- Commit the support needed.

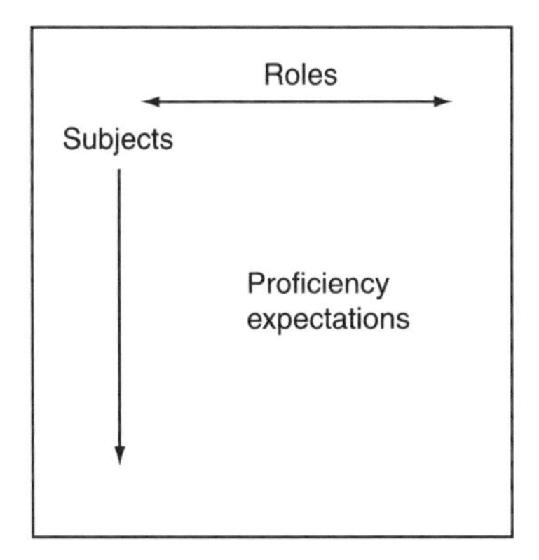

Figure 3.4 Defining required skills and expected proficiencies

addressing management questions, and go into the specific details of the work that will be required of management in creating and implementing the advanced learning model.

The emphasis of Figure 3.4 is that clear expectations must be set by management concerning the expected proficiencies of people within any given role in the organization. For each position a list of needed subjects must be generated. The expected proficiency of the person in each of the listed subjects must also be clearly defined. This must be done as a progression, or growth, of proficiency with seniority in the organization. It should take into account both employee career growth and anticipated future organizational needs. A second important point to emphasize with this slide is that once the subjects and expected proficiencies are

- Provide guidance in the form of expectations and recommendations.
- Assign expert teams to create the guiding documents.

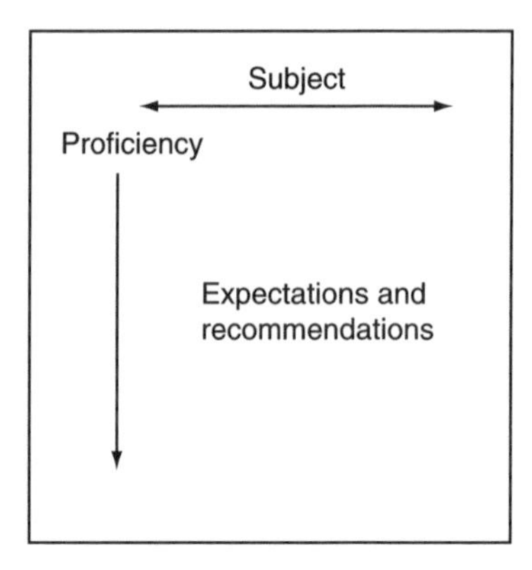

Figure 3.5 Provide direction on expectations and approaches

clearly defined each individual holds the primary responsibility for their own learning. Management, from the direct supervisor to the top management of the company, must commit to support each employee's continued learning.

Finally, the slide in Figure 3.5 suggests that management plays a further role in guiding the workforce toward appropriate development. The experience of expert practitioners, and the larger organizational context understood by management, lends guidance toward deciding both what subjects need to be learned, and how best to learn them. In other words, in addition to providing a clear set of expectations regarding the subjects and proficiencies required of each member of the workforce, management must provide guidance and recommendations to help the individual in seeking out the needed development.

In review, the purpose of this section was to provide the framework for a succinct presentation to a company's engineering management, intended to achieve their active support for the proposed advanced learning model. Of course one reason their support is needed is the fact that their direct involvement will be needed in the foundational steps of implementing this model. The next sections provide a more detailed look at what will be required in order to create and implement the advanced learning model.

3.3 Identifying types of positions

Fundamental to the advanced learning model is the need to provide a clearly defined set of subjects and proficiency expectations for each engineering position. In many industries one's immediate reaction may be that everyone's work is different. If this premise is accepted it may quickly undermine the implementation of this approach to employee development. The result could be a huge array of different positions, each having a long list of required subjects. Recognizing that the success of this system will rest on providing specific expectations and development recommendations for each subject, it is clear that the result will be a system that will quickly die of its own weight.

The first step will be to study the company and its technical organizations as a whole. The goal of this study should be to identify patterns and similarities between positions – based on similarities of subjects, especially at their most fundamental level. Distinctions are often made between different roles in a company based on specific tools used, or responsibilities performed. Our interest in advanced learning focuses on the underlying knowledge and capabilities that result in technical leadership in carrying out these responsibilities. Categories of job roles should be created based on these fundamental subjects.

As a specific example let us now look at Cummins, Inc. Cummins is a worldwide manufacturer of diesel and gas engines and power systems. They employ approximately 26,000 people, of whom between 1000 and 1500 are engineers involved in product development, application and service. While the majority of these engineers are located at the company's headquarters in Columbus, Indiana, USA, a significant number are located at engineering facilities in the UK and India, and smaller numbers are dispersed through facilities around the world.

As this process for advanced learning was implemented at Cummins the first difficulty encountered was exactly that described in the earlier paragraphs of this section – at first look, one would easily conclude that there were a great number of very different engineering positions. As an example, let's begin with the engineers responsible for the performance, fuel economy, and exhaust emissions development of the engines. Engineers involved in this general area include everyone from those doing quite basic combustion research to those involved in the fine-tuning and incremental improvement of current production engines. The full range of engineering positions pertaining to performance, fuel economy and emissions is listed below:

- Combustion research – Engineers advancing the fundamental knowledge pertaining to the company's products, and developing models and tools to support product development.
- Advanced Product Development – Engineers developing various technologies for implementation on future products, with no specific production plans.
- New Product Development – Engineers developing specific products over an approximate range of one to three years away from production.
- Current Product Engineering – Engineers responsible for ramping up production on new products, and the ongoing development of current production products.

- Application Engineering – Engineers responsible for tailoring current production and new products to meet the needs of specific applications.

Each of the above groupings of engineers could be further broken down into those who worked on diesel engines, natural gas engines, fuel injection systems and turbochargers. Additionally, there was debate concerning whether engineers who worked on the very large engines needed proficiency in the same areas as those who worked on much smaller engines. Finally, in the category of Application Engineering, an argument could be made that whether one worked with boats, trucks and buses, agricultural and construction equipment, or electric power generators, would also make a significant difference in the subjects required.

The first step toward implementing the advanced learning model was to review in some detail the various types of engineering positions described above, as well as the similarly vast array of positions in other disciplines, with an eye toward the common fundamental subjects upon which expertise in any of these positions was based. Ultimately, the various positions described above were quite successfully addressed with the single category of 'Performance Development Engineer.' In progressing through this book, it will become increasingly apparent as to why such general categories can be used.

Extending the example throughout the engineering workforce at Cummins, the following position categories were selected:

- Mechanical Development Engineer;
- Performance Development Engineer;
- Design Engineer;
- Reliability Engineer;
- Structural Analyst;
- Experimental Mechanics Engineer;
- Metallurgist;
- Material Scientist (Non-Metallurgical);
- Service Engineer.

Even with this relatively small number of categories, many of the specific subjects identified were quite similar across two or more of the categories. In some cases engineers found themselves holding responsibilities best characterized by a combination of the categories. Each of these findings will be further discussed later in the book.

In carrying out the task described in this section, there might be a very natural tendency to start small and demonstrate the model with one organization. Perhaps this organization could be used as a means of gauging reactions and measuring results before implementing the model across the board. However, this approach is not recommended. It is recommended that the effort start with the roles in which the greatest number of engineers are working. Paring down from the great number of roles to a small number of more general categories will rapidly result in a system that accounts for a great majority of positions. As the work progresses to account for smaller and smaller numbers of people the resulting categories will tend to stabilize, with the last few types of positions rather easily falling into existing categories or pairs of categories. If on the other hand, the process was initiated with a small group of specialists, and larger job descriptions were addressed later, the natural tendency would be to create more and more categories, which would then be more difficult to pare down.

As an example, at Cummins there are perhaps twenty engineers working in the field of materials science. Each engineer in this field tends to be quite specialized. If the process of identifying job categories was initiated here, arguments could easily be made for having separate job categories for metallurgists, ceramicists, tribologists, lubricant chemists, and so forth. Carrying the process forward into other organizations, the precedent would have been set for a wide variety of job categories. While it could be argued that the number of categories could later be reduced, it should be recognized that this would be a considerably more difficult process, especially if the engineers were already working with the system – and unless the entire system is implemented at once, they certainly will be. At Cummins, the system was implemented over a period of approximately three years.

In addition to the obvious reason of keeping the process manageable, there are several further reasons for keeping the number of position categories small. Each of these will become more apparent as the discussion continues, but they are introduced as follows:

1 A relatively small number of position categories will aid in providing focus and helping employees to identify job interests and development needs. It is much easier for an employee to think in terms of the general type of work in which she or he is interested, than to try to identify and work towards a very specific position.
2 A small number of position categories will be much more manageable in the organization's efforts toward succession planning. This will be discussed in more detail in Chapter 9.
3 Finally, the small number of positions will aid the organization in ensuring compensation equity across varieties of specific positions responsible for similar types of work.

3.4 Identifying required subjects

Once the job categories are agreed upon, the next task is to identify the specific subjects required of engineers working in any of the roles falling into that category. This is perhaps the most difficult task in the creation of the advanced learning model. The challenges fall into two areas. First is the issue of consistency across job categories. The second challenge will concern the question of how finely to break down the required subjects. Some engineers will have a tendency to create a long and detailed list of very specific subjects, while others may be content with a few coarse 'buckets' of related subjects. Creating a comfortable balance that everyone can work with will require some trial and error.

The first key to generating an accepted listing of subjects is to clearly place the responsibility for doing so with the technical experts within the discipline, under the leadership of a senior management representative. The Training Leader's role must be limited to that of the process facilitator. The subject listing should not be generated by one person, but by a small team of engineers recognized as leading practitioners in the particular position category. The engineers should be selected to represent a range of specific roles and job responsibilities from which the category was developed. Throughout this process it must be remembered that the engineers in the particular discipline, or job category, must feel strong ownership in the resulting subject list. They must feel comfortable that the list accurately reflects the underlying technical subjects they need to do their work.

Once the team is assembled the members must each become comfortable with the overall objectives of the project. The same material used to achieve management buy-in, as discussed at the beginning of this chapter, should be reviewed with this team. The team will now be at the point where they understand what they will need to generate. At this time they will, in all likelihood, enter into a discussion of the challenges identified at the beginning of this section. They will want to gain clarity on what is meant by 'subjects,' and they will want to determine how finely they should break down their list. Again, remembering that it is imperative that they 'own' the results, it will be important not to provide any rigid guidelines to address either of the challenges. For the proper functioning of this model it will not be necessary to have consistent definitions of subject areas. The subject lists for various job categories may in fact look quite different, as will be demonstrated later in this section.

Perhaps the best way to proceed is to now apply the 'brainstorming' approach. The team members should begin listing subject areas with very little discussion or critique, and without concerning themselves with consistency, in either terminology or breadth. This process will quickly result in a rather long list, from which the team can work. Once the initial list has been generated, further discussion will need to center on combining some terms, breaking others down further, and adding or eliminating still others. The process begins to get easier at this point, as everyone on the team will begin to gain comfort with the direction and results of the discussion. While it may take a couple of meetings, with opportunities to reflect on the process and discussion in between, an agreed list will usually come together quite naturally.

Throughout this section, I have called upon the Training Leader to resist the temptation to steer the process, again and again emphasizing the need for the engineering staff to own the results. At this time the point must be emphasized again, because once two or three such subject lists for different position categories have been created, it will probably be apparent that the teams have not been consistent in their approaches. One team may have used very academic distinctions between subjects, with each subject corresponding closely to a particular course or topic of the university engineering curriculum. Another team may have chosen categories that reflect specific tasks commonly carried out by engineers working in the roles being discussed. Still another may have chosen subjects that correspond closely with particular attributes of the product being developed. In order to further clarify these different approaches, partial skill lists from two of the job categories developed at Cummins, Inc. are given below:

Performance Development Engineer
- Fuel Consumption
- Transient Response
- Torque and Power
- Startability

Structural Analyst
- Stress Analysis
- Fatigue Analysis
- Linear Vibration
- Thermal Loading

Clearly, very different approaches to subject identification are being taken in the two position categories shown above. The 'Performance Development' team adopted a subject list based on specific attributes of the products being developed. A 'subject' based on one of these attributes involves the various fundamental concepts that must be understood to complete the task of optimizing the performance of the product in terms of that attribute. The 'Structural Analyst' team selected a subject list corresponding closely to the academic topics underlying their work.

It should be apparent that both of these approaches to generating a list of required subjects can be equally effective in clarifying the capabilities required of engineers working in the particular roles. It is therefore not important that a consistent approach be taken across the various position categories.

At this point in the process, a subject listing has been developed for the particular category of positions, and everyone on the team involved in creating the listing should feel personal ownership in the results, and a willingness to defend it and support its use in their organizations. It has been the author's experience that a subject listing of between about 15 and 25 specific technical subject areas is a very workable breakdown for most positions. While some engineers may prefer a finer breakdown, and a resulting greater number of subjects, this runs the risk of overly taxing the resulting system, thus wasting resources and undermining support. A shorter subject list may seem attractive, but if the subject categories remain too coarsely defined, it will be difficult to articulate learning needs and differentiate between the capabilities and experiences of various individuals. This will limit the effectiveness of the tool in supporting employee development.

3.5 Identifying required proficiencies

Chapter 2 concluded with an introduction to the topic of proficiency scales, and the need for some such scale as a means of articulating one's growth in proficiency. It will now be necessary to implement a consistently applied scale as a means of communicating learning expectations throughout the company. Referring back to Chapter 2, an alphabetical, numeric, or descriptive scale may be chosen. While scales with finer distinctions often seem appealing, they increase the complexity of the system, and really do nothing more than increase discrepancies of interpretation. A scale of no more than four or five categories is recommended, with the following approximate distinctions:

- little or no familiarity with the topic;
- familiar with the basic concepts and terminology;
- able to apply the concepts in commonly seen applications;
- able to apply the concepts in advanced applications, and teach others;
- recognized leading expert on the topic.

It should be noted that in the field of education, Bloom's taxonomy was intended to very carefully articulate growth in learning, providing a consistent set of terminology that could be used across a wide range of applications, from early childhood through to adulthood. Bloom's taxonomy consists of six levels, and could certainly be used for our purposes. Unfortunately, as will become apparent in the next chapter, the efforts required to complete the implementation of the proposed model increase significantly with each increase in the number of proficiency levels. A small number, focused on creating only the distinctions necessary to rapidly identify organizational capabilities and needs, is strongly recommended. In most cases, all that is needed is a means of identifying those who can relatively independently complete a required task, and that smaller group that have the further proficiency to address especially complex situations, and provide mentoring or instruction to others.

Once a proficiency scale has been adopted and consistently articulated, the task at hand will be to identify the desired proficiency levels of those holding any particular role in the organization. The result will be a 'learning matrix,' the general format of which is shown in Table 3.1.

Table 3.1 Learning matrix

Subject	Proficiency as a function of increasing job title →			
Subject A	2	2	3	4
Subject B	1	2	2	3
Subject C	1	2	3	4
Subject D	2	3	4	5
Subject E	2	2	3	3

3.6 Some specific examples

In this section three specific examples illustrating aspects of the use of learning matrices are presented. Two of these examples demonstrate the various uses to which the approach has been applied at Cummins, Inc. The third example is taken from the Society of Manufacturing Engineers, and their efforts to provide a standardized set of subject definitions and a standardized career development tool for their membership and others in the discipline of manufacturing engineering.

An example of a portion of an engineering learning matrix used at the Cummins Engine Company is shown in Table 3.2. It must be noted that in the work at Cummins the word 'skill' was used for what has been termed 'subject' in this book. The resulting matrix was referred to at Cummins as a 'skills matrix,' where the term 'learning matrix' is used in this book.

The particular example is for engineers working in mechanical design roles. The matrix uses a five-level proficiency scale, designated by the letters N, A, U, S and E. An 'N' designation identifies a subject that is **not needed** or expected of the engineer at the particular level of experience. The letter 'A' suggests a basic **awareness** of the subject – the individual would be expected to understand the basic concepts, terminology and definitions, but would not be expected to be able to independently apply the subject in their work. At the 'U,' or **user** proficiency the individual would be expected to be able to independently apply the subject under ordinary circumstances in their day-to-day work. The individual having a proficiency designated by the letter 'S' would be a **specialist**, able to apply the subject at an advanced level, and capable of teaching and mentoring others. Finally, the 'E' designation is used to signify an **expert** practitioner – one who is pushing the limits of understanding of the subject, and is recognized as a leading practitioner.

In the Cummins learning matrix the set of subjects identified for the particular grouping of positions is listed in the left-most column. Across the top is the progression of position titles corresponding to an increase in experience and salary grade. A dual progression path is reflected in this

Table 3.2 Example portion of engineering skills matrix

Skills Matrix – Mechanical Design

Skill	Engineer	Sr. Engr.	Group Ldr.	Tech. Spec.	Manager	Tech. Adv.	Chief Des.
Engineering fundamentals							
Stress analysis	A	A	A	A	A	U	A
Dynamic analysis	A	A	A	A	A	A	A
Fatigue analysis	A	A	A	A	A	A	A
Machine design	A	U	S	S	S	S	E
Reliability	A	A	A	A	A	A	U
Tribology	A	A	A	A	A	U	U
Noise	N	A	U	U	U	U	U
Vibration	A	A	A	A	A	U	U
Torsional vibration	A	A	A	A	A	A	A
Thermal loading	A	A	U	U	U	U	U
Ferrous metallurgy	A	A	U	U	U	U	U
Non-ferrous metallurgy	A	A	U	U	U	U	U
Gaskets/Seals/Elastomers	A	A	U	U	U	U	U
Bolted joints	N	A	U	U	U	U	U
FE analysis	A	A	U	U	U	U	U
Engine knowledge							
Engine components	A	U	U	U	U	S	E
Engine sub-systems	A	U	U	U	U	S	E

Used by permission: Cummins, Inc.

table. When an engineer progresses beyond the role of senior engineer, a managerial ('Group Leader') or technical ('Technical Specialist') title may be given, based on the primary responsibilities associated with an individual's role. The expected proficiency in each subject area is then provided versus the progression in experience and responsibility level.

NOTE: It is very important to emphasize that the learning matrix is not to be seen as a series of hurdles to pass in order to receive the next promotion. The matrix is intended to provide a guide for professional development. At Cummins the proficiency recommendations were set with the specific intention of encouraging each employee to stretch in their learning – they were not necessarily expected to be capable of performing in each subject area at the recommended level.

Another example from Cummins, Inc. is that of defining job classifications for various unionized positions. With the goal of creating a common language across various organizations and manufacturing plants, and reducing the number of job classifications, the classifications were defined by documenting the skill requirements of any given position. Such a resulting job classification is shown in Table 3.3. In this case a sentence description of each skill is given in the left-most column. The same proficiency scale described in the previous example makes up the right-most column. Additional columns are included to rank the criticality of each subject to the particular job classification, and to identify the expected time period over which the subject could be learned if the person does not already possess proficiency in the subject when moving into the particular role.

The Professional Competency System developed by the Society of Manufacturing Engineers is summarized in Figure 3.6. The Career Map shown centrally in the figure is analogous to the learning matrix developed in this chapter. In this case the subjects are broken down in a multi-level Competency Catalog, and defined in sufficient detail to accurately support a proficiency assessment process. 'Competency' in this example is used synonymously with 'proficiency' in our discussion. The assessment process, and the importance of accurate self-assessment, will be discussed in Chapter 5. The Assessment leads directly to the development of a Diagnostic Report and Personal Development Plan. The SME Professional Competency System is further described in the publication by Wright and Tillman (2000).

Table 3.3 Example job analysis identifying capabilities required by job classification

Key responsibility	Importance to job			Time to learn					Competency level		
*(Importance: **N**ot Critical, **I**mportant, **C**ritical) (Time to Learn: **0** = None, **1** = Brief (0–2 wks), **2** = Intermediate (6 mths), **3** = Extensive (6 mths +), **4** = Continual) (Competency Level: **A**wareness, **U**ser, **S**pecialist)*											
SPECIALTY WORK (Work required of Associate – Publications)	N	I	C	0	1	2	3	4	A	U	S
Responsible for marketing, generating and implementing publishing type work. This work may include combinations of the following:			C					4			S
As approved by management, maintain contact with outside design firm/personnel on an as-needed basis and manage these projects; confer with authors.		I				2				U	
Assist in expediting project workflow by developing and maintaining expertise in the company-standard applications for text and visual communication.		I				2				U	
Prepare written material for business, technical and electronic publication, including reading copy to detect errors in spelling, punctuation and syntax; mark with standard proofreading symbols; rewrite or modify copy to conform to standard publication criteria and corporate identity and guidelines; verify facts and clarify information; organize material and select appropriate page layouts and font conforming to publication standards and corporate requirements.			C			2					S

Manage process of creating, updating and maintaining web sites for intranet and internet; provide changes and updates to the web pages of customers' sites using the web content management tools.		C	1				S
Work with customers to conceptualize, design and create visual concepts; ensure appearance of projects.		C			4		S
Develop business graphics and creative/complex design work from a variety of verbal and written inputs; develop themes and secure related material.	I				4		S
As appropriate, select or recommend graphics, such as drawings, diagrams, pictures and charts, select and crop photographs and illustrative materials to conform to space and subject matter requirements.	I			2		U	
Research, analyze, develop and create electronic layouts, text and artwork for print and web-based publishing.	I		1			U	
Resolve problems concerned with developing and publishing material; provide customer assistance through problem solving.		C			4		S
Review final proofs with customers and make revisions as required.	I		1			U	

Used by permission: Cummins, Inc.

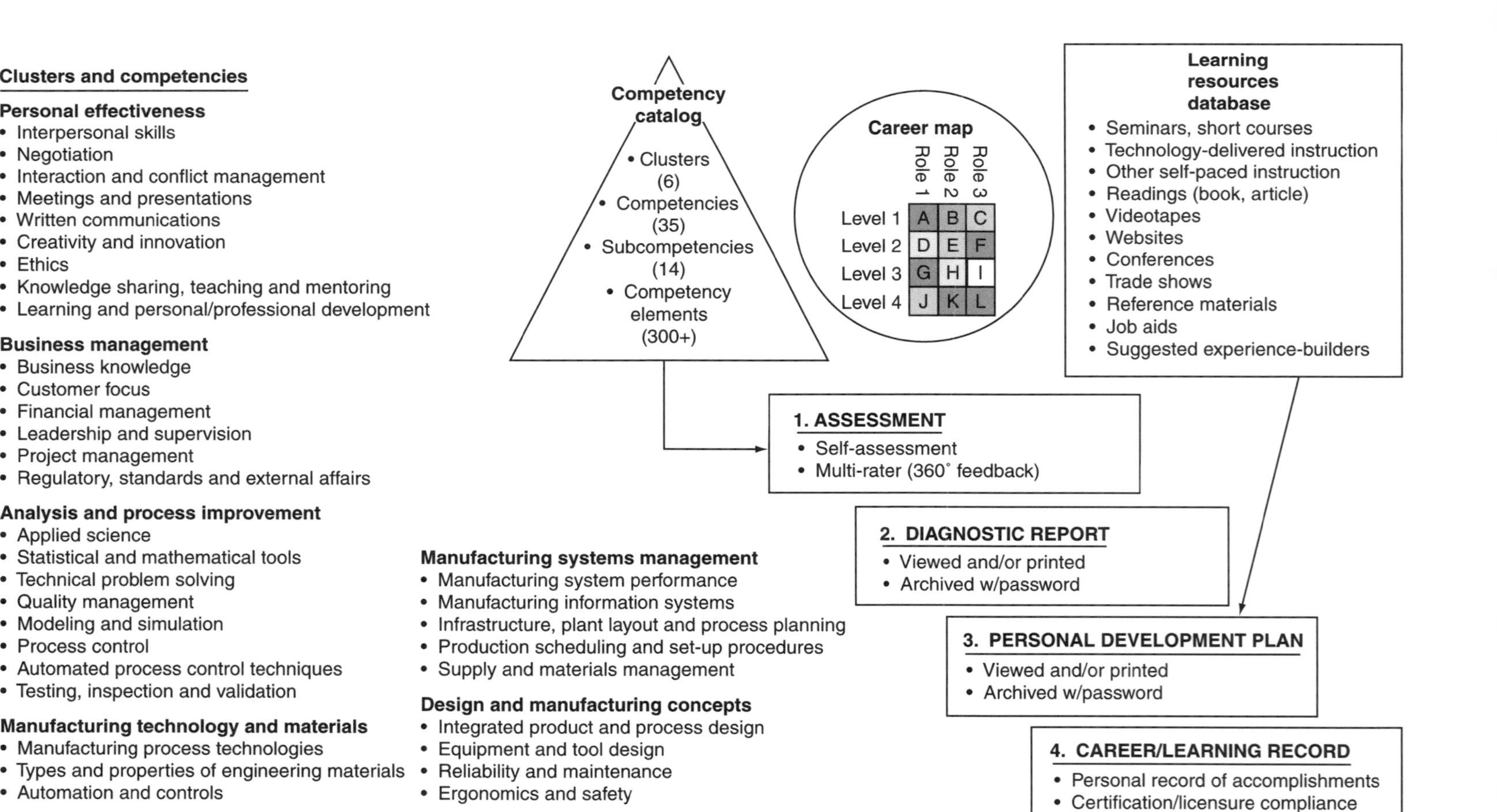

Figure 3.6 SME professional competency system

3.7 Summary

- Recognizing that the needs of the technical workforce are not well served by the approaches most often emphasized in corporate training activities, a new learning model will be required.
- Because the required approach to technical learning in the workplace is very different from today's employee training approaches, the necessary shift in thinking must begin with top management support.
- The specific expectations this new learning model places on top management are to:
  - set expectations regarding the subjects and proficiencies required in each engineering role;
  - provide non-threatening, developmental tracking mechanisms;
  - clarify the expectations placed on the engineering workforce and their direct supervisors;
  - communicate recommendations and guidance for developing proficiency in particular subjects.
- A suggested summary of the key points that must be covered in order to achieve management support was summarized in Figures 3.1 to 3.5.
- The learning model presented here is based on a learning matrix, requiring the following ingredients. (The first three ingredients were discussed in this chapter. The fourth ingredient will be discussed in Chapter 4.)
  - A relatively small number of types of roles in which each employee will be able to identify their own work.
  - A listing of the subjects required for performing each role.
  - The proficiencies expected of employees in each role at various levels within the organization.
  - Practical recommendations for developing proficiency in each subject.

Chapter 4

Mechanisms for advanced learning

In Chapter 2, several practical scenarios were introduced where an industrial need called for a training intervention. The final such scenario demonstrated the case in an engineering organization where mechanisms were needed to support the development of advanced proficiencies in a small number of employees. This example served as a springboard for the topic of Chapter 3 – involving management in clearly defining a progression of expected proficiencies. The resulting learning matrices, and proficiency expectations, provide a framework of career development goals for all technical employees. It was also suggested in Chapter 3 that management must then take the responsibility for providing guidance to their employees on where to turn for the requisite advanced learning. The purpose of this chapter is to take a detailed look at how new learning is acquired, and the types of learning interventions appropriate for supporting development of advanced proficiencies.

As has been emphasized since Chapter 1, the continuing education of a company's technical workforce must focus on the development of advanced proficiencies and capabilities. It is here that a company must support technical employee development efforts in order to remain viable in an environment of rapid technological advance and increasing international competition. Let's now look again at the example from Chapter 2:

A critical aspect of a particular company's product durability is ensuring against fatigue failure under certain commonly seen operating conditions. There is a small group of engineers dedicated to this problem, and they have developed some very effective techniques. However, one of their top people has recently retired and another will retire in two years. At the same time, in an effort to reduce product development time and better meet local customer needs, more and more of the development work is occurring at plant sites around the world. There have recently been product failures attributed to the fact that these development groups have not applied the appropriate techniques.

In this example, a few people, located in a particular division of the company, have developed expertise in areas important to the success of the company's current or future product. However, they are continually faced with projects that take priority over developing training material, and strategic decisions made by the company's management have placed many of the people needing their expertise at distant geographic locations. The problem is exacerbated by the fact that they are beginning to lose some of their knowledge base through retirements. The challenge facing the company's development and training department is to develop programs that will effectively transfer the working knowledge of these experts to the larger number of practitioners (as was indicated in Figure 3.1) over a reasonable period of time.

4.1 Methods of attaining the higher proficiencies

The challenge facing the development and training department in the example is not trivial. It is presumed that these engineers already have a solid background in the particular discipline (fatigue analysis, in the example). The employee development program needs to build on the fundamental knowledge they possess, and help them increase their capabilities to apply their learning – both in commonly seen situations and under special circumstances. Referring again to Table 2.1, the goal of such continuing education must be to move from Bloom's second level (Comprehension) to the third level (Application), the fourth level (Analysis), and possibly the fifth level (Synthesis). Clearly no single training program would foster this entire progression of proficiency development, but the company's vision may very well need to include a series of programs and experiences that would allow such a progression over time.

It is no coincidence that most adult training tends to focus on the more basic proficiencies. The next step the technical teams formed in Chapter 3 must take is to provide guidance that engineers can use in planning for their continued professional development. It is almost certain that the immediate focus of these teams will be on developing training programs. The next challenge that must be faced will therefore be to encourage the teams to look beyond training programs, and to encourage them to think about how they themselves learned the skills they want their junior colleagues to gain.

As one plans programs to address increasingly advanced proficiencies the training methodologies become much less well-defined and less agreed upon among educators. West and his colleagues (1991) make a distinction between **well-defined**, **moderately well-defined**, and **ill-defined** problems in education. They describe a well-defined problem as one for which there is a single, widely agreed-upon way of understanding and dealing with the educational need. A moderately well-defined problem is described as one for which there are two or more acceptable alternatives to select from, and a widely agreed-upon process for making the selection. In contrast, they describe an ill-defined problem as one for which no such prior guidance has been established. Based on these definitions, the development of advanced proficiencies in engineering education most clearly fits the category of an ill-defined problem, and we may well be left with more questions than answers as to how to proceed. The next section is intended to provide a theoretical basis that can be shared with the engineering teams to help them think about alternative, experientially-based, skill development mechanisms.

4.2 Fundamentals of experiential learning

Referring back to Figure 2.3, as learning progresses from an understanding of basic concepts into increasingly sophisticated application, the learning occurs more and more through experience. Extension of learning programs to support the development of higher proficiencies will thus necessarily focus on experiential approaches. In this section the fundamental precepts of experiential learning will be explored before identifying specific methods applicable to advanced technical learning. Kolb (1984) presents a four-component model describing the required steps in effective experiential learning. The first step is to get the learner fully and openly engaged in new experiences. This is followed by creating the environment in which the learners can work with one another, under the guidance of the instructor, to reflect on and interpret the new experiences from multiple perspectives. The third step in Kolb's model involves building on the individual backgrounds of the learners to logically integrate these new experiences with their prior capabilities. Finally, the resulting new capabilities are exercised in making decisions and solving problems. As experiential elements are incorporated in program development this model serves as an excellent guide toward ensuring the effectiveness of the program.

Central to experiential learning is the active involvement of the learner in the learning process. This is consistent with many studies of adult education that emphasize active involvement in learning; the work of Knowles (1980) is often cited in this regard. The role of the learner is nicely summarized by Caffarella and Barnett (1994) as follows:

> Adults can call upon their past experiences and prior knowledge in formulating learning activities, as well as serve as resources for each other during learning events. Experiential learning activities such as reflective journals, critical incidents, and portfolio development, can provide opportunities to introduce adult learners' past and current experiences into the context of learning events.

Jackson and MacIsaac (1994) identify two elements enhancing the effectiveness of experiential learning. The first again emphasizes active learning as they state the role of the learner to 'not simply absorb knowledge but actively construct it.' This 'constructivism' involves the use of prior knowledge in interpreting, retaining, and revising new information. The second element identified by these authors is the importance of matching the instructional context as closely as possible to that of its application in the field.

4.3 Practical approaches to experiential learning

Based on the fundamental concepts of experiential learning summarized in the preceding paragraphs, specific experiential methods applicable to technical skill development will now be discussed. Lee and Caffarella (1994) present an extensive list of methods for engaging learners in experiential learning – both within a classroom program, and in the field. An abridged listing is given in this section, with comments concerning the use of each method in technical education. While it is beyond the scope of this book, the variety of preferred learning styles should be kept in mind as these methods are discussed – more than one method may be appropriate in support of any given subject.

4.3.1 Demonstration with a return demonstration

An analytical or experimental technique is demonstrated, followed by an opportunity for the learner to repeat the demonstration. This method of experiential learning can be adapted for classroom use, or incorporated in on-the-job learning. In making recommendations for advanced skill development, a specific sequence of observation followed by guided personal application could be called out. An example of such a skill development recommendation might be:

> This experience might be gained by requesting a demonstration of the technique by a specialist from Department X. Then ask the specialist to observe and guide you as you repeat the technique. It is recommended that this technique be learned in application to your specific project work, but the same process may be applied under a hypothetical, or practice situation, in preparation for an expected, upcoming need.

4.3.2 In-class case study

A practical problem is presented, learner discussion and questions are encouraged, and a solution technique is presented. There may be situations where your team representing a particular discipline feels strongly that a subject requiring experiential learning is best taught in the controlled environment of a classroom. This is one of various techniques that can be used to increase the experiential element of classroom learning. The following is an example of how this approach might be implemented:

In order to help you develop this capability, seminar Z has been created. In this seminar, led by specialists from Department X, actual situations experienced by our company will be presented. You will be given the opportunity to work with a small group of colleagues, and under the guidance of your class leaders, in applying this advanced solution technique to the case study problem. To register for this course . . .

4.3.3 Critical incidents

In the context of an engineering problem this technique would involve having class participants present specific problems they have faced, and using these practical problems to discuss solution techniques. This is a variation on the approach discussed in the previous section. It may be appropriate in some situations, but is limited by the fact that most engineering problems must be addressed immediately, and cannot be put off until the next class session. An example of how this approach might be recommended is:

In order to help you develop this capability, seminar Z has been created. In this seminar, led by specialists from Department X, you will be asked to bring to the class an actual problem you are responsible for solving. You will be given the opportunity to work with a small group of colleagues, and under the guidance of your class leaders, in applying this advanced solution technique to your problem and those brought by others in the class. To register for this course . . .

4.3.4 Poster presentations

In programs where it is recognized that various participants already have experience and expertise that would be helpful to the development of others, poster presentations provide a forum for sharing and discussion. Such sessions could periodically be arranged within a company to enhance communication of solution techniques, and encourage the interaction and communication by which learning is effectively transferred. This approach may be applied and recommended as described here:

Twice each year our company holds a poster session in which specialists from various divisions display and demonstrate the techniques they have successfully applied in solving problem Q. As a way of gaining proficiency with these techniques it is recommended that you attend an upcoming session. The schedule and location are posted at . . .

4.3.5 Case study research

After the demonstration of applying a new analytical or experimental technique the learners complete a specific assignment to analyze and reflect upon the demonstration. This approach may be used to further the effectiveness of the case study method described earlier. After working through a case study in a classroom the individual is expected to apply the technique in one's own work, under the guidance of a specialist – perhaps an instructor from the case study class. An example of how this recommendation might be used is:

After completing seminar Z you will be expected to apply the techniques of the seminar to one of your project assignments. Arrangements will be made for the seminar instructor to work with you on this application. It is strongly recommended that the application be made within six months of completion of the seminar.

4.3.6 Trips and tours

This method could involve a visit to a site where a product is used, or a visit to a laboratory where a new technique is being utilized. It may be formally incorporated into an organized professional development event, or it may simply be recommended as something to be done individually. Examples of how either approach might be recommended are as follows:

Arrangements are in place for visits to the ACME laboratories where you can meet with one of the expert technicians and participate in a hands-on demonstration of the new X test procedure. Call 124–4657 to schedule your visit.

In order to learn more about how the ABC product is used by our customers you are encouraged to visit one of the company's field service and support centers. Employee visitors are welcome at any time, and you will have the opportunity to talk directly with customers about their experiences with the product.

4.3.7 Coaching

The learner performs a new analytical or experimental technique while being observed by, and receiving feedback from, an expert in the application of the technique. This approach allows direct, hands-on learning to occur in an environment where the risk of error is minimized, and the individual can obtain immediate answers to questions. The learning occurs very rapidly with a minimal investment of resources. If such opportunities are formalized, the inefficiencies of individually seeking out guidance are minimized. An example of how such an instructional technique can be catalogued follows below:

It is recommended that this technique be learned at the time at which it can be directly applied to your project. Call the EXP department at 234–5678, and arrange to meet with one of their technicians. You will be given the opportunity to carry out the procedure under step-by-step guidance, and your questions will be addressed as they arise.

4.3.8 Mentoring

As the learner performs tasks incorporating new learning, a formal assignment has been made for someone possessing expertise in the tasks to act as a resource, and to discuss and review findings. While similar to the **Coaching** technique described in the previous section, **Mentoring** differs from coaching in at least two important ways. First, the mentoring relationship is ongoing. An individual is assigned a mentor who will provide guidance on one or more topics over a period of time. Second, the mentor does not typically provide step-by-step guidance as does the coach. The mentor is a resource to be called upon as needed. In some situations meetings between the individual and the mentor may be scheduled to occur on a regular basis, while in other cases such meetings occur informally, or 'as needed.' A sample paragraph recommending mentoring as a development technique is given here:

Developing expertise in the field of XYZ requires practical application of the techniques over a variety of problems. The development of your proficiency will be greatly enhanced by working with a mentor with whom you can arrange to periodically review your progress and learning. The following people have agreed to provide mentoring on the specific topics listed after their names. Please contact them directly to arrange for their help and guidance on your projects.

4.3.9 On-the-job training

The learners are assigned tasks that build on their past experience and incorporate the opportunity to apply new learning. This important learning method will be discussed in much greater detail in Chapter 7, where the supervisor's role in employee development is taken up. At this point the topic can be summarized by stating that selecting project assignments that build on one's current proficiency while simultaneously calling on the person to learn new subjects is an extremely effective approach to professional development. Any statements recommending learning through specific project assignments will encourage this employee development technique.

4.3.10 Clinics

An opportunity is arranged for learners to present their application of what they learned to a panel of experts, in a non-threatening critique and discussion forum. While this technique is seldom used in industrial settings, it can be very effective. One approach might be to use the clinic in conjunction with a classroom training program. Individuals or teams within the class are given an assignment to complete a project. A team of experts is brought into the classroom to then review the projects, hear from the learners, and provide feedback to them. Using the technique in a classroom setting makes it far less intimidating than if the same technique was applied in the review of someone's day-to-day work. A specific example, recommending a class where a clinic is included, might be as follows:

In order to learn the processes of widget design it is recommended that you attend the class, 'Fundamentals of Widget Design,' offered once per quarter by our engineering training center. In this class you will learn the step-by-step process of widget design, and how to best evaluate and optimize the design for a particular application. You will then team up with two or three other class members to produce your own widget to meet a set of hypothetical criteria. A team of widget design experts will then visit the class, review your designs, and provide helpful feedback and recommendations based on their collective experience.

4.3.11 People networking

Arrangements are made for learners to periodically meet and review the application of what they learned with one another, each thereby learning from the applications experienced by others. This technique is often used quite effectively as a follow-up activity some period of time after a classroom training program. However, it can also be organized as a standalone opportunity. Arrangements can be made for forums, where people from around the company working on similar problems can periodically meet and share their findings and learning with one another. This can be done under the guidance of a facilitator or a forum leader having expertise on the topic, or it can be done without such leadership. A skill development recommendation for participation in such a discussion network follows:

Forums on the topics of EFG are held the second Tuesday of every month, at 9:00 AM, in the Fireside Lounge. These peer discussion groups are open to anyone working on the development of EFG, and provide an opportunity for everyone to share what they've learned and build on the learning of one another. Advance registration is not required, and the discussion agenda is decided upon by the participants at the beginning of each forum.

4.4 Putting it all together

In Chapter 3 the technical leadership was called upon to develop learning matrices that spell out proficiency expectations, and to provide recommendations as to how an employee might best develop the expected proficiencies. The chapter concluded with several examples of the resulting learning matrices. So far in Chapter 4 the focus has been on a necessary shift in the way learning is viewed – in order to develop the needed advanced proficiencies, the emphasis must shift from classroom to experiential learning. A number of examples of approaches to explicitly encourage or direct experiential learning were then provided.

The work that remains is to create the specific, experience-based recommendations for developing advanced proficiencies in each subject. In other words, for each subject identified in the matrices of Chapter 3, a set of recommendations must be provided, guiding the employees in their individual development toward increasing proficiency. These recommendations may be provided in an outline or paragraph format. General examples of what these recommendations might look like are provided in paragraph format in Figures 4.1 and 4.2. In Figure 4.1 a very simple skill that should be familiar to every reader is described in order to demonstrate the kind of information that should be included. Figure 4.2 shows the same approach applied to a far more sophisticated engineering skill. The same examples are repeated in outline format in Figures 4.3 and 4.4 respectively. A close comparison of Figure 4.2 with Figure 4.4 reveals that it is very difficult to convey the same level of detail when

Preparing canned soup

Proficiency Level 1

At this proficiency the individual will be expected to be able to identify the required ingredients and utensils. Under the supervision of someone experienced in preparing canned soup, the individual will be capable of completing the steps of preparation.

This proficiency can be gained by studying the instruction label found on the soup can. Mentoring is widely available.

Proficiency Level 2

At a Level 2 proficiency the individual is expected to be able to independently prepare a can of soup, following the instructions posted on the can.

This proficiency is best gained through experience, under the guidance of a mentor.

Proficiency Level 3

This proficiency requires the individual to be capable of modifying the canned soup recipe for improved flavor. The individual should also be able to make a can of soup stretch to feed 50 percent more hungry people, while still maintaining an appealing flavor, again by carefully modifying the recipe and adding ingredients as necessary.

This proficiency can best be gained by trial and error. Discussions with others who have had experience making similar recipe modifications will also be effective.

Figure 4.1 Proficiency expectations for common task, in paragraph format

Stress analysis

Proficiency Level 1

The engineer should be able to review the design and operation of a given component, identify the primary loads, load paths and necessary assumptions, and apply the fundamental equations of mechanics to obtain an estimate of design integrity. The engineer should also be able to determine an appropriate computational or experimental approach for obtaining more detailed analyses of the design.

An undergraduate engineering elective in stress analysis or mechanics of materials provides the required grounding. Guidance on specific problems is available through the company's Structural Analysis group.

Proficiency Level 2

The engineer should be able to critique a design to identify potential failure locations, produce calculations to estimate the likelihood of failure, and be able to propose design improvements and demonstrate their effectiveness through calculation. The engineer will be expected to be able to make appropriate assumptions, then select and utilize a model of sufficient but not excessive accuracy and complexity from available computational techniques. The engineer should also be able to identify simple experiments or measurements that can be used to test computational results or provide key information for improving the analysis.

This proficiency is achieved through demonstrated experience in application of stress analysis principles to a variety of design problems, and should include at least basic skills in infinite element analysis.

Proficiency Level 3

At this proficiency a key characteristic is experience and demonstrated success at applying the equations of mechanics in conjunction with simple experiments to solve structural problems. The engineer should also be able to apply the equations of mechanics to special cases such as elastic/plastic problems, three-dimensional load paths, and combinations of thermal and mechanical loading.

Several years of experience in the application of stress analysis principles to product design and development, over a wide range of components and problems, is required. Application of stress analysis should be supplemented by graduate course work.

Figure 4.2 Proficiency expectations for an engineering subject, in paragraph format

using the outline format. Wording must be chosen very carefully to avoid questions and misinterpretation. This will be important in employee self-assessment, and in ensuring consistent documentation of employee proficiencies across the corporation. These topics will be discussed further in Chapters 6 and 9 respectively. It is the author's experience that, given the choice between the two formats, most engineers will prefer the full paragraphs. While the outlines might seem easier to prepare, the required attention to wording makes them quite similar in required effort to the paragraph format.

Preparing canned soup

Proficiency Level 1

Expectations:

- Identify the required ingredients and utensils.
- Complete steps of preparation with supervision.

Development recommendations:

- Study the instruction label found on the soup can.
- Mentoring is widely available.

Proficiency Level 2

Expectations:

- Independently prepare a can of soup.

Development recommendations:

- Guidance of a mentor.

Proficiency Level 3

Expectations:

- Modify the canned soup recipe for improved flavor.
- Make can stretch by 50 percent while maintaining appealing flavor.

Development recommendations:

- Trial and error.
- Discussions with experienced soup modifiers.

Figure 4.3 Proficiency expectations for common task, in outline format

An example from the advanced learning tool as implemented at Cummins, Inc. is given in Figure 4.5. The approach taken at Cummins was first to provide paragraph descriptions to help the employee understand the expectations associated with a particular proficiency level, and then provide recommendations regarding how to achieve each proficiency. The initial paragraphs, supplementing the basic proficiency scale used on the skills matrix, were very effective in allowing accurate self-assessment of an employee's current proficiency. Spelling out what was expected at each proficiency left little room for debate concerning the current capabilities and competencies of the individual, and went a long way toward removing assessment conflicts between the individuals and their supervisors.

4.5 Summary

- This chapter began with a discussion of the need for experiential learning mechanisms. The importance of experiential learning was established in Chapter 2, and is key to the learning model being presented.

Stress analysis

Proficiency Level 1

Expectations:

- Review design and operation of a given component.
- Identify primary loads, load paths, necessary assumptions.
- Apply fundamental equations of mechanics.
- Obtain estimate of design integrity.
- Determine appropriate computational and experimental approaches.

Development recommendations:

- Undergraduate engineering elective in stress analysis.
- Guidance on specific problems from Structural Analysis group.

Proficiency Level 2

Expectations:

- Critique a design to identify potential failure locations.
- Produce calculations to estimate the likelihood of failure.
- Propose design improvements – demonstrate effectiveness.
- Make appropriate assumptions, and select computational techniques.
- Identify simple experiments or measurements.

Development recommendations:

- Experience in variety of design problems.
- Basic skill development in finite element analysis.

Proficiency Level 3

Expectations:

- Experience and demonstrated success solving structural problems.
- Apply principles to special cases such as:
  - elastic/plastic problems;
  - three-dimensional load paths;
  - combinations of thermal and mechanical loading.

Development recommendations:

- Several years of experience over a wide range of problems.
- Supplement with graduate course work.

Figure 4.4 Proficiency expectations for an engineering subject, in outline format

- The following practical tools for experiential learning were introduced, and brief examples of each were given:
  - demonstration with a return demonstration;
  - in-class case studies;
  - critical incidents;
  - poster presentations;
  - case study research;
  - trips and tours;

Engine cooling systems

Cooling systems familiarity begins with a general understanding of the cooling circuit, and the major cooling system components. Cooling fluids, filtration and additives should be understood. Water pump operation, performance and efficiency, along with water pump drives and seals should be familiar. Cooling jacket design, including critical cooling regions and design criteria, and venting must be known. Knowledge of thermostat design and operation, oil cooler and charge air cooler design is also required. Vehicle installation considerations such as fan and shroud design, radiator sizing, vehicle coolant requirements (cab heaters, etc.) and deaeration must also be understood. Component tests for water pumps and heat exchangers, flow visualization tests of cooling jackets, and rig tests for deaeration and system performance (flows and pressures) should be familiar. Test cell measurement practices for heat rejection measurements should also be known.

Skill development progression

Awareness – Familiarity with cooling systems operation and components, coupled with application to cooling system problems.

An Awareness proficiency can be attained through exposure to cooling system design problems, coupled with participation in training seminars. The Fundamentals of Reciprocating Engine Design seminars include a unit on cooling systems. The Pilot Installation Center at the Technical Center offers resources and expertise in vehicle installations. The Heat & Fluids group in Cummins R&D organization provides expertise in water pump design, cooling jacket design and component testing.

User – Awareness proficiency plus experience on projects pertaining to various aspects of cooling systems.

The User proficiency is attained through involvement in cooling systems projects addressing the various aspects of the system. Simultaneous skill development in the areas of Flow Circuit Analysis and Application Engineering is recommended. Skill development may be further supplemented through additional training seminars. The University Consortium for Continuing Education offers a two-day short course on Vehicle Engine Cooling Systems. Call (818) 995-6335 for additional details. The University of Michigan offers a three-day short course each summer in Ann Arbor, entitled Flow Visualization Techniques with Image Processing. Call (313) 764-8490 for further information.

Specialist – User proficiency plus experience leading cooling system design – both engine jackets/circuit and vehicle installation.

It is expected that the specialist will have led the design of each aspect of the cooling system, as described in the opening paragraph, from conceptual analysis through testing and validation, and into production. The engineer must possess personal technical experience in water pump design and performance, cooling jacket design, circuit analysis, venting, filling and deaeration, heat rejection measurement and heat exchanger sizing.

Figure 4.5 Example of a technical subject description, with proficiency expectations

  - coaching;
  - mentoring;
  - on-the-job training;
  - clinics;
  - people networking.

- The final step in preparing the framework for this experiential learning model is to create recommendations that employees can use to develop advanced proficiencies. These recommendations should be based on one or more of the experiential learning tools just listed.

Chapter 5

Communicating the information

In previous chapters we have discussed an approach by which engineering management can create a framework encouraging advanced and continuous learning in the workplace. The creation of a learning matrix providing a listing of subjects and recommended proficiencies for each engineering position was presented in Chapter 3. In Chapter 4 some specific techniques for advanced learning were described, and paragraph or 'bullet point' recommendations for how to develop advanced proficiencies were demonstrated.

Based on the work of Chapters 3 and 4 the company will now have a series of learning matrices similar to that shown in Table 3.2, and a write-up or outline of expectations and recommended proficiency development methods for each subject listed on any of the matrices. These write-ups will be of the types shown in Figures 4.1 to 4.5.

We are now faced with the question of how and where to maintain this information such that it will be easily updated and available to all of the employees throughout the company who have need of it. The purpose of this chapter is to discuss this question in further detail, and recommend electronic techniques for maintaining and distributing the information.

5.1 Needs of the communication system

Only a few short years ago the only viable option for disseminating the information being discussed would have been to create notebooks containing the learning matrices, expectations and development recommendations. An elaborate indexing and page numbering scheme would have been required, and a process would have been needed to ensure that updated pages were regularly distributed to all notebook holders, with instructions to remove the old pages and insert the new. A scheme for making the updating process fail safe would become a requirement of quality criteria such as those demanded by ISO 9000 certification. Such an approach would be very unwieldy indeed, and would have required compromises concerning how widely the information could be distributed – one notebook per department perhaps.

With the advent of **hypertext** software, which became commonly available in the early 1990s, a tool has been created that lends itself extremely well to addressing the problem at hand. Hypertext software takes advantage of the computer's capability to store and retrieve data in a non-linear fashion. Rather than stepping through data in a single path or progression (analogous to a written book), the computer allows files to be 'linked' in whatever ways might be specified by the programmer.

The application of such linking to the learning matrices and individual subject files (proficiency development recommendations) is portrayed in the diagram of Figure 5.1. Each 'page' or electronic file is shown as a box in the figure, with the chosen links between these pages shown as lines. The individual reading the document selects particular pages to access by activating the desired link with the computer's mouse. Referring again to Figure 5.1, the user enters the database at an introduction screen. The user is welcomed to the database, and may receive various on-line instructions or other information on how the database is to be used to support learning within the particular company's system. Upon reading the introductory material the user activates the first link. In the structure shown in Figure 5.1 the user is given only one choice at this point – a link to a page where the various positions and available learning matrices are

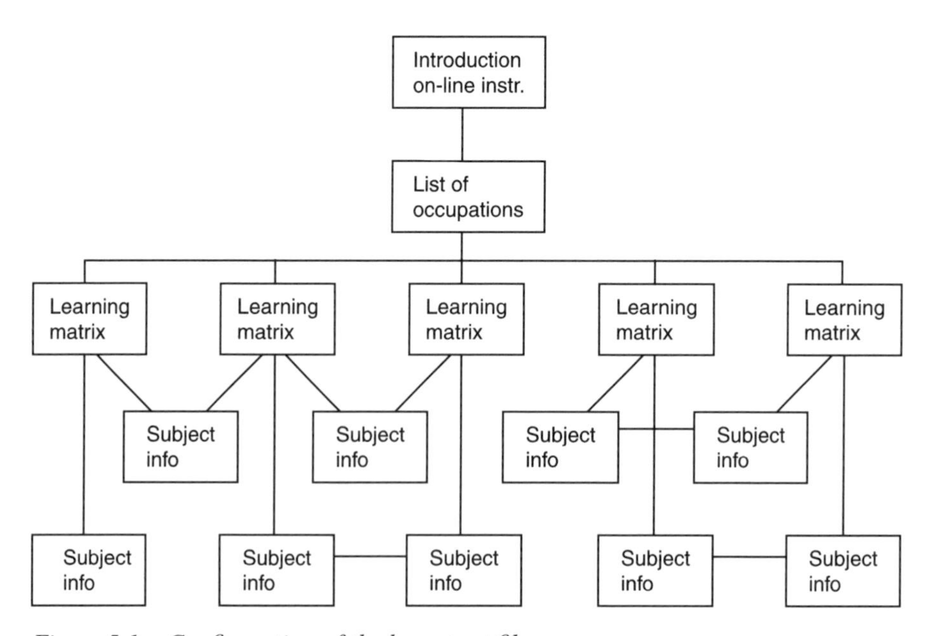

Figure 5.1 Configuration of the hypertext files

listed. The user may then select the position of interest, and activate the corresponding link.

The next screen that appears is a particular learning matrix, showing the expected proficiencies as one progresses in the organization. The real attraction of the hypertext software now becomes apparent – each of the subjects listed in the matrix serves as a link to the file describing that subject, setting forth expectations regarding the practice of the subject, and providing recommendations for proficiency development. It should be apparent from this discussion that the file structure described here is exactly that which gives the internet its power and capability. The internet is one of the earliest and most widespread uses of hypertext software and is in fact a very attractive environment for the construction of the learning matrix database.

Before proceeding further in discussing the construction and operation of the learning matrix database it will be instructive to review the features that the database must possess. Some of these have been introduced already, but for completeness they are listed in the sections below.

5.1.1 Common subject definitions

Quite often it will be found that the same subject is referred to in two or more learning matrices. Engineers working on different projects, under different job titles or position descriptions, and perhaps in different parts of the company, may be required to possess differing accumulations of subject expertise. However, a portion of the subjects required in any particular position may very well be shared with the subject requirements of individuals in other positions. It will certainly be in the company's best interests to arrive at a common definition of any subject used in positions throughout the company. An obvious advantage is in avoiding duplicating the efforts involved in creating the subject definition, expectations at each proficiency, and recommendations for development. Another advantage follows directly from the resulting common terminology and expectations. In Chapter 9 we will discuss how this database we've created can be utilized in resource planning throughout the company, and in rapidly and effectively addressing the specific subjects needed on a project. With such a database the company would be in a position to conduct an internal search to identify all people possessing expertise in a particular set of subjects at some specified proficiency level. If a common and accurate definition of the subject has been used, and a measurable set of criteria has been described for identifying one's proficiency level, an objective approach is then available for finding all possible people within the company qualified to fill the particular position. A person from another location, not previously known by the department having the open position, might be identified. The common subject definitions will provide reasonable confidence that the person identified possesses the expertise to fulfill the requirements of the position. While this will not substitute for an interview process and further evaluation, it will identify people who may otherwise have been overlooked. There are strong and obvious implications regarding issues of workforce diversity as well – the 'playing field' has been leveled, and **all** potentially qualified people have been identified, as opposed to just the ones with whom the department manager has recently enjoyed a round of golf!

The hypertext file linking system lends itself very well to this need for common subject definitions. A single file is maintained for each subject, and this same file is accessed by each of the learning matrices calling out the particular subject. The file is 'owned' by a particular organization possessing expertise in that subject area. They are responsible for revising the information (definition, expectations as proficiency is achieved and development recommendations), and keeping the most up-to-date information available in the database.

5.1.2 Linkage between matrices and subject write-ups

This topic is closely related to the one just described, and is directly addressed by the hypertext linking capability. The subject list maintained in any matrix contains a hot link from each subject to its supporting documentation. Each time a subject file is updated in any way the new file is automatically made available to each of the matrices specifying the subject area. The revised file must be given the exact same file name as its predecessor. In other words, it completely replaces the earlier file – the file name cannot include a revision number or date. If it is important to a particular company to maintain a historical database on changes to the information, the old files would need to be archived with their names changed in some way. It should be noted here that it may be important to include revision dates on each of the files. In the author's experience, including a revision date and the name of the group within the company responsible for the information in a given subject file was very beneficial in ensuring the regular updates and reviews specified for ISO certification.

5.1.3 Write control

It should now be apparent that the company incorporating this learning system will want to carefully create not only the information that goes into the database, but a system of responsibility for its maintenance. While the information needs to be available to the entire engineering workforce, the authority to update or change the information must be carefully designated. One would certainly not want an individual engineer to be able to change any portion with which she or he disagrees.

Again, hypertext programming quite naturally addresses this issue. With virtually any hypertext software environment, the 'read-only' software is made available to everyone – most often as free, or 'share' ware. An internet web browser is an example of hypertext read-only software. Write privileges can be separately specified, and the documents protected against unauthorized modification. Using the internet as an example, web editing software is required to create or modify files. While web editing programs are becoming widely available, further protection against unauthorized changes comes about by placing the files on a secured server. Read access may be permitted, while passwords and special permission are required to open the files in an editor.

5.1.4 Revision and document control

This requirement follows directly from what has been discussed in previous sections. Because of the rapid pace with which technology is changing, employee development in engineering and science is an extremely dynamic process. It will therefore be important to ensure that the employee development information is maintained in such a way as to always reflect best practices and the state-of-the-art at any snapshot in time. At the beginning of the chapter the difficulty of maintaining a printed book of employee development information was briefly discussed. Each time an update was made a process was needed to ensure that outdated pages were removed and destroyed, and replaced with the latest revisions. At the very least, such a process is extremely cumbersome and it certainly creates the opportunity for error. Again, the hypertext electronic database provides a very attractive solution. A single file, at a single location, is maintained, containing the information pertaining to each subject. When updates are required, the file is modified, and the modified file replaces its predecessor. From that point in time onward, whenever anyone accesses the information pertaining to that particular skill they automatically receive the updated version. It was previously recommended that a revision date be included within the content of the file; by doing so it provides a check for those who decide to print out any of the subject information, allowing them to ensure that they are working from the latest revision level.

5.1.5 Ease of use

While it may be quite humbling to say so, the company's training director must always acknowledge that personal development is seldom the number one day-to-day priority of the employee. Almost all employees will recognize their long-term needs for development, but very few can afford the luxury of giving such development high priority in their day-to-day work. Recognizing this sad fact, the training director must then admit that even the smallest additional burden or required processes placed upon a person in support of his or her development will have the potential of bringing the system to its knees. Many a well-intentioned employee development system has quickly succumbed and died from its own weight. Perhaps the single most important lesson to be gained is that an effective learning system must be one that can be utilized nearly without effort by its 'customers.' It is imperative that anything supplied to the employees in support of their development must require virtually no learning – its use should immediately be intuitive to anyone for whom it has been created. They should be able to begin using it without training; they should be able to jump on and off during the few minutes between phone calls or meetings, and be able to get what they need during those few minutes. Admittedly no learning system known to humankind will be able to function solely in this manner, and the role of management coercion will still have its place. It will be apparent in the remaining chapters that, while the learning system presented in this book significantly reduces resource requirements, it has not fully eliminated them – that will have to be the subject of a later book! Nevertheless, what ever can be done to minimize the time required in learning and applying a new system – especially one seen by most employees as peripheral to their work – should certainly be sought. The software required for this database will certainly be a key to this objective. Utilizing software tools already familiar to most people will contribute greatly to the ultimate success of the database. Several options are available as a framework for the hypertext database. Wide familiarity with the internet suggests it may be an obvious choice. This will be discussed further in the next section.

5.1.6 Accessibility

Finally, the more accessible we can make the database, the more useful it will be to the workforce for whom it is designed. The electronic format allows it to be accessed from wherever a networked computer is available. In the technical workforce today, the electronic format will in most instances allow the database to be accessed from each individual's desktop. This does, however, require that the desktop computers be linked to a common server – either through a local area network (LAN) or modem connection.

5.2 Hypertext database software

At the beginning of this chapter hypertext software was described as a relatively recent family of programs developed to take advantage of the computer's capability to organize information in ways not possible with the printed page. Most of us are now using hypertext software, though often without identifying it as such. The internet has previously been mentioned. The programming language of the internet is referred to as **html**, an acronym standing for **h**yper**t**ext **m**arkup **l**anguage. The 'Help' files associated with most 'Windows' program applications are also examples of hypertext databases. In either case, the use of words, phrases, or 'buttons' enabled as hot links create the ability to navigate through files in a way selected by the user, based on a file structure defined by the programmer. Early in the development of such software various hypertext authoring programs were made commercially available. With each of these programs, the purchase of an authoring, or writing program license then included a compatible reader that could be supplied as shareware to each user. Such programs could very effectively be used to create the electronic database presented in this book. The only real disadvantage to such programs is that each one has its own learning curve and would require a format and a reader program not already being used within the company, or by the employees for whom the database is being supplied.

At the time of writing one of two possible approaches will be most likely to already be in wide use by any given company's technical workforce. One of these is the internet. Many companies now provide internet access to their employees and that number is growing each day. While the internet is a public access network, companies can create 'firewalls' blocking access to files created within the company, and thus using the same language and file management system for their own 'intranets.' Many companies are now using Lotus Notes for their internal communications. This powerful software tool also utilizes the hypertext file linking format, and was developed specifically to provide enhanced file security, and sophisticated file management capabilities needed for many business and engineering activities. For companies using Lotus Notes this provides a second attractive framework for the learning matrix database.

This chapter concludes with a series of figures showing examples of the learning matrix database as implemented at Cummins Engine Company, Inc. Note again that the Cummins system shown here used the term 'skill' in the same context as this book is using 'subject.' The 'learning matrices' discussed in this book are referred to as 'skills matrices' in the Cummins example. In this case the database is provided within the company firewall on an intranet fileserver. Figure 5.2 shows the introductory screen, as opened with a commercial web browser. This initial screen provides welcoming information and several links to optional instruction pages further describing the learning system, and how the database fits in with other elements of the employee development processes specific to Cummins. This information will be most beneficial to first-time users, so its viewing is optional, and the user can immediately scroll down to the learning matrix listing, the top of which is visible in the figure. From this page any one of the matrices created for the company's engineering and

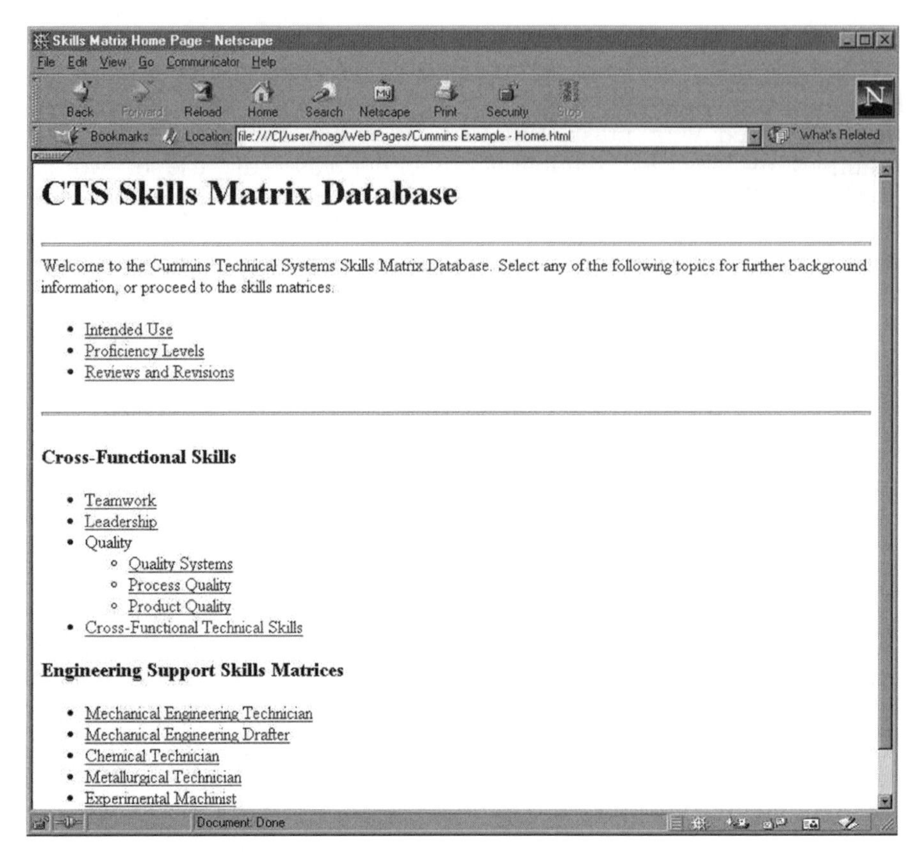

Figure 5.2 Cummins' skills matrix home page

engineering support workforce can be accessed. Note that each position description is an active link to another file. Clicking on any of these links brings up the screen showing the particular learning matrix. Such a matrix is shown in Figure 5.3. The matrix contains more subjects than will fit on the screen, so as indicated at the right-hand side, the user can scroll up and down to view the entire matrix. Each of the subjects listed in the matrix provides a further link. 'Clicking' on any of these subjects opens the file containing the subject definition, expectations versus proficiency and recommendations for proficiency development. An example of a subject information screen as viewed with the web browser is shown in Figure 5.4.

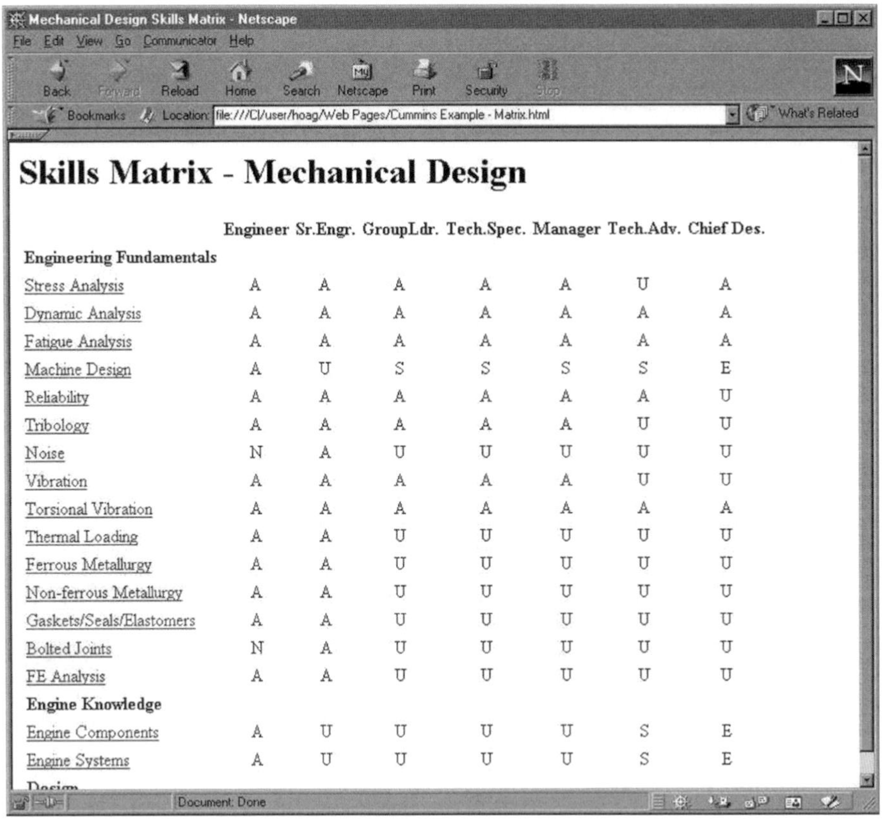

Mechanical Design Skills Matrix - Netscape

Skills Matrix - Mechanical Design

	Engineer	Sr.Engr.	GroupLdr.	Tech.Spec.	Manager	Tech.Adv.	Chief Des.
Engineering Fundamentals							
Stress Analysis	A	A	A	A	A	U	A
Dynamic Analysis	A	A	A	A	A	A	A
Fatigue Analysis	A	A	A	A	A	A	A
Machine Design	A	U	S	S	S	S	E
Reliability	A	A	A	A	A	A	U
Tribology	A	A	A	A	A	U	U
Noise	N	A	U	U	U	U	U
Vibration	A	A	A	A	A	U	U
Torsional Vibration	A	A	A	A	A	A	A
Thermal Loading	A	A	U	U	U	U	U
Ferrous Metallurgy	A	A	U	U	U	U	U
Non-ferrous Metallurgy	A	A	U	U	U	U	U
Gaskets/Seals/Elastomers	A	A	U	U	U	U	U
Bolted Joints	N	A	U	U	U	U	U
FE Analysis	A	A	U	U	U	U	U
Engine Knowledge							
Engine Components	A	U	U	U	U	S	E
Engine Systems	A	U	U	U	U	S	E

Figure 5.3 Cummins' skills matrix example

5.3 Review of where we've been

We have now completed our discussion of the creation and communication of the learning matrix database. At this point the tool is available to our technical workforce. In the chapters that follow, the use of this learning system will be discussed from several perspectives. We will begin by looking at our new learning system from the perspective of the individual engineer – how she or he will be expected to use it, and what they can gain from doing so. The role of the supervisor will then be discussed, and we will begin shifting our attention to the corporate systems that are complemented by the system, or that can be developed around the system. The direct use of the system in employee development will be covered, followed by a few further benefits made possible by the system.

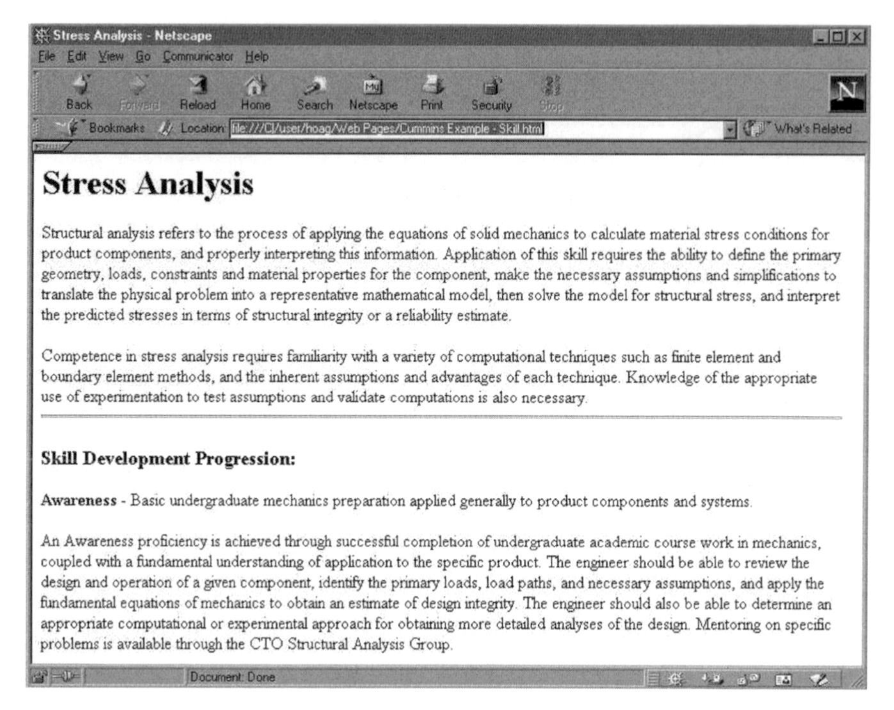

Figure 5.4 Cummins' skills description example

5.4 Summary

- The focus of this chapter was on the creation of a database for maintaining and sharing the learning matrices and proficiency development recommendations.
- The database must meet the following criteria:
    - common subject definitions;
    - linkage between learning matrices and subject definitions, recognizing that more than one learning matrix may call upon the same subject;
    - write control;
    - revision and document control;
    - ease of use;
    - accessibility.
- Hypertext software (such as an intranet, or Windows 'Help' files) is very well suited to meet the criteria required of the learning matrix database.
- This chapter concluded the construction and implementation of the experiential learning model. The next chapters will discuss its use in the workplace.

Chapter 6

Employees own their own development

In this chapter and the next the learning matrix database developed and implemented in the preceding chapters will be applied to the individual employee's continued development. This chapter discusses employee learning from the perspective of the individual employee. In the next chapter the roles and expectations of the employee's direct supervisor will be taken up.

The perspective taken in this chapter begins with the assumption that **each person holds primary responsibility for their own professional development**. The notion that employees can or should sit back and wait for 'the company' (in the form of the supervisor, organization, or training department) to take the lead in telling the employees what training they need, and when, is to be completely rejected. This is not intended to sound harsh, but to emphasize that it is clearly in the employees' best interests to take the lead in their own development. The company's role is to encourage and support each person in their learning process. Because employees do not always recognize this fact, the company needs to provide a clear message to this effect. The learning matrix database is a tool to provide aid and guidance, and set expectations pertaining to the employee's efforts.

This chapter begins with the employees' review of the database to select a learning matrix that best reflects the specific work they perform, or if necessary, to combine skills from two or more matrices. Successive sections will then cover the employee's self-evaluation concerning their current proficiencies in each subject area, and the prioritization of development needs based on current responsibilities and future needs and interests. The chapter will conclude by introducing the role the supervisor must play in implementing an employee development plan, setting the stage for the further discussion of the supervisor's roles in Chapter 7.

6.1. Learning matrix selection and modification

When the initial list of positions was created in Chapter 3, it was emphasized that not all positions could be covered, and that in some cases an individual may have to create a personal matrix from two or more of the general matrices. Using a specific learning matrix, particular subjects may be emphasized or de-emphasized based on the work requirements of a particular position. In many instances the individual may work with a 'subset' of subjects within a matrix. In other instances subsets of subjects from two or more matrices may be combined to create the particular repertoire of subjects needed for the person's role.

In many organizations there are nearly as many different engineering roles as there are engineers in the company. In other words, no two engineers perform exactly the same type of work. In Chapter 3 the need to resist the temptation to create a separate learning matrix for every possible engineering position was emphasized. Instead we identified several general position types, and tried to ensure that the resulting 'buckets' would reasonably closely match the positions of the majority of engineers within the company.

Now, as we make the resulting database available to the engineering workforce, the engineers must review what is being offered and 'find' their positions – their individual roles within it. In most cases this will be quite straightforward. One engineer's role might be to design a new widget; another's might be to redesign the current wangle. A general learning matrix has been provided for 'Design Engineers,' and both of these engineers recognize this matrix as pertinent to their roles. Once they begin reviewing the specific subjects within the matrix they may very well find that their specific roles require greater or lesser emphasis on individual subjects than that set forth in the general recommendations.

In other cases the subjects required in a particular role might be those found in a combination of two or more matrices. The engineer must review the matrices for the various types of positions representing elements of the work for which she or he is responsible, and select subjects from each matrix that are most relevant to the particular position. As an example of the various ways in which learning matrices might be applied to individual roles, let us take the case of a company that produces home heating and air conditioning products. The following learning matrices have been developed:

- Design Engineer.
- Structural Engineer.
- Thermal Science Engineer.
- Application Engineer.
- Service Engineer.

6.1.1 Case 1

An individual engineer has responsibility for the design of a new combustor for a natural gas furnace. Upon reviewing the learning matrix for a Thermal Science Engineer she finds that this matrix includes several subjects pertaining to the performance and sizing of refrigeration and air conditioning components, but also includes fundamental subject areas pertaining to energy balances and flame temperature calculation, and applied subjects in the areas of combustion, exhaust emissions and nozzle design. These latter subjects are quite relevant to her role and should be included in her personal inventory. Similarly, she may find subjects in the Design and Structural matrices that are also relevant to her assignment. Some of the same subjects may have been listed in more than one matrix, perhaps with different recommended proficiencies. In this example, subjects pertaining to thermal loads may have been identified in both the learning matrix for the Thermal Science Engineer and that for the Structural Engineer. It might have been recommended that the Thermal Science Engineer be familiar with the basic issues and terminology, while the Structural Engineer is expected to be able to perform detailed predictions and analyses of component temperatures and resulting thermal stresses. The engineer designing the new combustor may need to conduct the thermal stress analysis herself, or may be expected to work with an expert in that particular field who will lead this portion of the project. The proficiency she will need in this subject area will depend on these expectations.

6.1.2 Case 2

A plant several hundred miles away from the corporate headquarters and technical center produces air conditioner compressors. At this plant an engineer with several years' experience in compressor design has recently been assigned the role of team leader for a small group of engineers responsible for the ongoing development and application of the compressor produced there. The team includes individuals possessing backgrounds in compressor reliability and durability, compressor performance development and compressor application. While the team leader can rely on the expertise of each member of the team to contribute to the projects they receive, his own effectiveness will be enhanced as he gains greater knowledge in each of these engineering disciplines. In creating a personal inventory of required subjects – including both his current strengths and the areas in which his development might focus – he will need to review a variety of learning matrices. Subjects relevant to air conditioning compressor development might be identified from each of the matrices. The proficiencies he will need in each area will most likely be only basic familiarity, although a particular area might be selected for more advanced development, based on current project needs, or his personal future career goals.

6.2 Self-appraisal

Once the employees have identified the repertoires of subjects most relevant to their individual positions they are then called upon to self-evaluate their current proficiencies. This term, 'self-evaluation,' may very well generate skepticism among many an engineering manager. It is quite natural to expect that engineers will paint very rosy self-portraits – especially if there has been any hint (stated or unstated) that future promotions, raises, or attractive project assignments will be in any way dependent upon the subjects and proficiencies they can claim. And of course, a well-implemented learning matrix database provides the foundation to do exactly as the engineers might expect. Nevertheless, the use of self-evaluation is extremely important for several reasons, and as will be discussed later in this section, management fears of this process are unfounded.

First, let us look at why the self-evaluation process is so important. The following paragraphs present three reasons for formalizing self-evaluation as the foundation of the ongoing employee development process.

6.2.1 Ensuring that the process is non-threatening

It is an unfortunate fact of life in most corporations that employee development is perceived as part and parcel of employee performance. Any critique of current employee proficiencies, and identification of further learning needs, is all too often interpreted, rightly or wrongly, in punitive terms. Providing clear, unambiguous definitions of the expectations goes a long way toward alleviating the natural concerns of the employee. This eliminates, or at least minimizes, the subjectivity normally inherent in proficiency evaluation. Allowing people to review their own abilities relative to this defined scale, and self-identify their proficiencies, will make most employees much more comfortable with the process. Most employees will be much more supportive of this process than they would be if the assessment of their capabilities was to be accomplished by others, who may or may not (more often not) be fully aware of the person's abilities.

6.2.2 Emphasizing personal ownership

This chapter began with the emphasis on each employee's ownership of their own professional development. Self-evaluation follows naturally from this strategic decision. Failure to incorporate self-evaluation is inconsistent with personal ownership in the process, and undermines the culture of continuous learning that the company should be instilling.

6.2.3 Minimizing management time requirements

Finally, an important pragmatic argument stems from the many demands already placed on the time of any company's engineering managers. If self-evaluation is not accepted as the starting point, supervisors will need to be called upon to evaluate the proficiencies of every engineer in their organization. Depending on the other demands they are facing, and on the importance they place on this process, they may or may not give this process the attention required for an accurate assessment. Even if they treat the process quite seriously, it should be clear that they will not be able to conduct as accurate an assessment as an individual could provide of themselves, unless they have worked with the employee for a long time, over a wide variety of projects. The individual employee will inherently place greater importance on the process, and be able to more quickly produce a more accurate picture.

With these arguments for the importance of self-evaluation, I will now return to my earlier statement that concerns about the use of self-evaluation are unfounded. The key element that will in fact make self-evaluation successful is the care with which the proficiency expectations are written. A well-written statement of the specific tasks an employee will be expected to perform at a given proficiency will leave very little room for questions – the employee either possesses particular experience or does not. As the employee conducts the self-evaluation she or he will certainly wish to claim as high a proficiency as possible over a wide variety of relevant subjects. However, this must be tempered by the realization that if a particular expertise is claimed the employee may very soon be called upon to demonstrate exactly that expertise. The embarrassment of failing at this future task provides strong incentive for accurate self-reporting. As time goes on, and the procedures recommended here are utilized for the employee's professional development, the employee will gain further comfort with the tool, and see the career development advantages inherent in accurate reporting – they will receive the development opportunities they most need, and will receive project assignments providing the best balance of building on their current proficiencies and supporting the development of new ones.

When the learning matrix database was first implemented at the Cummins Engine Company, the concept of self-appraisal received considerable attention. Several engineering leaders were quite vocal in their opposition to this approach. In order to address their concerns, and ensure the validity of the self-evaluation, a pilot study was conducted. A group of about 120 engineers, from locations around the world, was asked to participate in a study where they were each to evaluate their proficiencies in several subject areas, followed by similar rankings conducted by their supervisors, and one or two peers who work closely with them. Using the five-level scale described in Chapter 3 this study found three-way agreement in just over 84 percent of the assessments done. The remaining 16 percent of the assessments were approximately equally divided between self-evaluations that were judged to have overstated and understated their proficiencies by one level. There were no instances of disagreements of more than a single level on the proficiency scale. This study was deemed to have demonstrated excellent results concerning the accuracy of the self-appraisal process. It is interesting, but perhaps not surprising, to note that the vast majority of those who overstated their proficiencies were very young engineers – especially those possessing advanced engineering degrees. Those most likely to understate their proficiencies were the most senior staff members, and those judged as having the least personal insecurity about their value and contribution to the organization.

In many organizations a simple form or spreadsheet might be desired as a way to ensure a standardized process, or possibly to allow further electronic assessment. The form might be as simple as a column listing subjects, followed by columns indicating desired and current proficiencies. An example of such a form is shown in Figure 6.1. By using a numerical proficiency scale and an electronic spreadsheet the form could be used to calculate a discrepancy between current and desired proficiency, and use this difference as one factor in determining development priorities. A graphical indicator, or the selective use of color, results in a visual measure of development across a department; the posting of these charts is attractive in many organizations. Two examples of such tracking charts are shown in Tables 6.1 and 6.2. While these two approaches are significantly different, it should be apparent that both succinctly provide valuable information for department learning efforts. Use of the chart shown in Table 6.1, in which the names of each employee are given, must be considered very carefully; employees will often be uncomfortable with such public reporting. This type of chart should only be used in organizations already very comfortable with the employee development policies. While not using individual names, the chart shown in Table 6.2 also provides valuable information, and does so in a much less threatening way. Finally, the information could be posted to a relational database, and centrally maintained as a corporate tracking resource. While such systems are far more involved than is required to meet our current objectives, they result in nearly limitless possibilities for organizational assessment, planning and tracking of capabilities and their locations within the organization.

It should be emphasized that even the simplest skill tracking forms with which this discussion began very nicely meet the requirements for

Position: ____________________

Proficiency scale
1 = Unfamiliar
2 = Basic familiarity
3 = Working capability
4 = Advanced capability
5 = Expert

Subject:	Recommended proficiency	Current proficiency	Development priority

Employee signature ____________________

Supervisor approval ____________________

Date ____________

Figure 6.1 Employee proficiency tracking form

Table 6.1 Example approach to departmental skill tracking I

Subject	Mary	Samuel	Jack	Jen	Debbie	**John**	**Peter**	Lynn	**Andrew**	Anne	Sara
Thermodynamics	2	1	3	4	3	5	2	3	1	3	3
Fluid mechanics	3	2	3	5	4	4	2	3	2	3	3
Gas dynamics	2	1	2	5	2	4	1	2	1	2	3
Heat transfer	2	2	3	4	3	4	2	3	1	3	3
Vibrations	4	3	4	2	5	3	2	3	1	2	4
Fatigue analysis	4	3	3	3	4	3	1	2	1	2	4
Dynamics	4	2	4	3	5	3	2	3	3	3	3
Metallurgy	3	3	3	2	3	2	1	2	2	1	2
Design of exper.	4	5	4	3	3	4	3	5	2	4	3
Process of control	3	5	3	1	3	2	2	5	2	3	2
Machining	3	4	3	1	2	1	1	4	1	3	2
Quality systems	2	5	4	2	4	2	2	5	2	5	4
Release process	2	4	3	2	3	2	1	5	2	4	3
Customer needs	3	3	5	3	1	3	2	4	2	5	5
Application engr.	4	2	5	3	1	3	2	4	1	5	4

Mentors or teachers; areas of departmental expertise	Development needs; areas of departmental need	**Retirees**	**New Hires**

Table 6.2 Example approach to departmental skill tracking II

Subject	1	2	3	4	5
Thermodynamics	++	++	+++++	+	+
Fluid mechanics		+++	+++++	++	+
Gas dynamics	+++	+++++	+	+	+
Heat transfer	+	+++	+++++	++	
Vibrations	+	+++	+++	+++	+
Fatigue analysis	++	++	++++	+++	
Dynamics		++	++++	+++	
Metallurgy	++	+++++	++++		
Design of exper.		+	++++	++++	++
Process control	+	++++	++++		++
Machining	++++	++	+++	++	
Quality systems		+++++		+++	+++
Release process	+	++++	+++	++	+
Customer needs	+	++	++++	+	+++
Application engr.	++	++	++	+++	++

ISO certification in the area of employee training. Completing a handwritten document each year, signed and dated by the employee and supervisor, and maintaining a record of past documents (and thus a development history) is sufficient to completely satisfy the ISO requirements.

While I have chosen to devote a separate chapter to the supervisor's role, it must be noted here that the work just described is not done totally in isolation. The employee would be expected to create an initial draft and review it with the supervisor. The next step, in which learning needs are prioritized, must be based on consensus between the employee and supervisor concerning the employee's current proficiencies.

6.3 Prioritization of development needs

Once the self-appraisal has been completed and reviewed with the supervisor, the next step will be to prioritize the learning needs. Development priorities should be based on not only the discrepancies between current and desired proficiencies, but the following additional criteria as well:

- current project needs;
- anticipated, near-future project needs;
- employee's short-term and long-term career development interests.

In other words, the following questions need to be addressed:

1 What will the employee need to learn in order to contribute most effectively in the current and upcoming project roles?

2 What does the employee need to begin learning to prepare for personal future career goals?

While the highest priority must be placed on the subjects that are currently, or soon will be, needed in the employee's current assignment, longer-term needs and employee career interests must be identified as well, and initial steps toward addressing those needs should be taken. The following example will serve to illustrate this process and will be carried forward into the next sections.

Janet is a team leader at an automobile manufacturer, responsible for developing the lubrication system for the company's new engine. Over the past year Janet and her team have completed the design of most of the system components. In the coming months they will receive 'production-like' prototypes of these components, and will be conducting performance validation tests. Janet's long-term career goals are to move from her product development role into the Research and Development organization, where she would like to conduct computational and experimental studies of fluid mechanics and lubrication. Toward this end she is actively pursuing a masters degree through an evening program at a nearby university.

Janet created a list of personally relevant technical subjects from matrices supplied by the company for Design Engineer, Engine Development Engineer and Thermal Sciences Engineer. She has completed the self-evaluation of her proficiencies, as shown in Table 6.3. In the next section we will return to Janet's continuing development as we take up the subject of individual development planning.

6.4 Individual development planning

Once the employee has done a self-assessment of current proficiencies (and this assessment has been agreed upon by the supervisor) an individual learning or development plan can be created. In most instances one year is an appropriate window for development planning, with agreed progress checks at set intervals.

In Section 6.3 it was pointed out that the development plan should take into account both the current and anticipated needs of the project(s) on which the engineer works, and the individual employee needs to prepare for long-range career objectives. Based on these criteria the employee and supervisor must work together to prioritize the employee's learning needs. These priorities might be categorized as high, medium and low, with the intent of addressing all of the high priority needs, and some of the medium priority needs.

The recommendations for employee development, included in the learning matrix database, now serve as a tool for the employee and supervisor, aiding them in identifying approaches for addressing the

Table 6.3 Janet's current learning matrix

Position: *Team Leader – Lubrication System*

Proficiency Scale
1 = Unfamiliar
2 = Basic familiarity
3 = Working capability
4 = Advanced capability
5 = Expert

Subject	Recommended proficiency	Current proficiency	Development priority
Foundational science:			
Fluid mechanics	3	3	
Heat transfer	3	3	
Tribology	3	2	Low
Lubricant chemistry	2	3	
Engine components:			
Lube pump design	4	5	
Bearing design	3	3	
Heat exchangers	3	2	Medium
Oil pan design	4	4	
Experimentation:			
Flow measurement	4	5	
Temperature measurement	3	3	
Engine start-up testing	4	4	
Lube pump performance test	4	5	
Analysis:			
Flow circuit analysis – Steady	4	3	High
Flow circuit analysis – Transient	3	2	High
Hydrodynamic bearing analysis	3	2	Low
Computational fluid dynamics	2	2	Medium

needed learning. As emphasized in Chapter 4, the resulting individual development plan may include formal training, but should also include various other learning mechanisms. A heavy emphasis will be placed on on-the-job learning – building on the employee's current expertise and stretching through the selected project assignments to acquire new proficiencies.

At this point we will return to the example of Janet, the Team Leader developing an engine lubrication system at an automobile manufacturer. Her current learning matrix was previously presented, in Table 6.3. In the final column in this figure, a development priority was placed adjacent to each of the subjects where her identified proficiency is less than that recommended in the learning matrix database. Her work this year will focus on the validation of the lubrication system she and her team have been designing. All of her requisite subjects pertaining to experimentation are already at or above the recommended proficiency. However, the validation process will include some analytical studies of the flow circuit – matching measured data to a system model and then using the model to test conditions that are difficult to simulate experimentally. Janet has had little experience with these models and will need to gain this proficiency. While the company may offer a training seminar on the use of these models, most of her learning will occur as she applies the models to her project, perhaps with some guidance from an expert user of the tool.

In concluding our review of this example several further points should be emphasized. First, while Janet's current proficiencies are below those recommended in a few other areas (Tribology, Heat exchanger design and Hydrodynamic bearing analysis) these have been given low priority. This decision was made because she will not need these subjects in her current or anticipated project assignments.

While she currently holds the recommended proficiency in Computational Fluid dynamics, she and her supervisor have identified this area as having a medium development priority. They have done so because, while she does not need to further develop this proficiency for her current role, it is important for her future career interests. Janet and her supervisor will consciously look for ways in which she can learn more in this area during the coming year. Perhaps a course at the local university will help address this need, or perhaps she could contribute to another project requiring such analysis.

Finally, it should be noted that Janet possesses advanced proficiencies in lube pump design, performance testing and flow measurement. In fact she has worked on a wide variety of projects in this area and is her company's leading expert on lube pump development. Her development plan highlights this fact and points to a role she can play in the development of others. This will be important both in helping other employees develop the proficiencies they need and in ensuring that Janet is not held back from the career moves she desires. As this example demonstrates, the use of this simple but well-documented development plan serves to identify where resources exist to aid in the development of others, and can very nicely address the often frustrating situation where an employee cannot make a desired transfer because proficiencies critical to the organization are not possessed by others within the organization.

6.5 Summary

- The fundamental point of this chapter is that each employee holds the primary responsibility for their own professional development.

- Company management has a responsibility to provide guidance and support (as defined in Chapter 3). The direct supervisor also plays an important role (to be discussed in Chapter 7).
- In order to participate in the learning model presented here each employee must identify the learning matrix, or possibly combination of two or more matrices, that best characterize their work.
- The subject definitions and expectations versus proficiency, developed as discussed in Chapter 4, provide a basis for very accurate employee self-assessments. Skeptics are encouraged to re-read Section 6.2, where the following advantages of self-assessment are discussed:
    - ensures that the process is non-threatening;
    - emphasizes personal ownership;
    - minimizes management time requirements;
    - often results in greater accuracy.
- The resulting skills assessment and tracking process has been demonstrated to meet ISO 9000 requirements, and is a far simpler system to implement than the more typical training tracking processes.
- The self-appraisal of employee proficiency over the range of subjects needed on the job will identify 'gaps' between current and desired proficiency that must then be prioritized based on:
    - current project needs;
    - anticipated future project needs;
    - individual career goals.

Chapter 7

The supervisor's role

This chapter addresses two very different aspects of management involvement in the employees' continued learning process. The first three sections closely parallel Sections 6.2 to 6.4 of the preceding chapter. In Sections 7.1 and 7.2 we will very briefly revisit the self-assessment and employee development prioritization processes, now from the supervisor's perspective. We will spend a bit more time on creating the development plan in Section 7.3, as it is here that the supervisor plays an especially critical role.

In Section 7.4 we will look at another aspect of supervisory support. This aspect is that of corporate-wide support; the employee's direct supervisor may play a role in this aspect, but this role will be shared with the technical leadership throughout the company.

7.1 Supervisor review of self-assessment

As was emphasized in the previous chapter, the self-assessment process must be 'owned' by each individual employee. The supervisor's role is one of cross-checking the process, ensuring that any differences of view get resolved, and then providing approval and support. The supervisor's work should be concentrated on any areas of disagreement. Conscientiously conducting this work reveals a couple of further benefits of the proficiency assessment process.

Recognizing once again that a well-written description of expected capabilities at each level of the chosen proficiency scale will go a long way to achieve an accurate assessment, the supervisor can typically expect to be in quite close agreement with the employee's self-assessment. In most cases there will be very few areas in which the supervisor might have concerns about the self-assessment. An important role of the supervisor is to take up each of these areas of disagreement in a one-on-one discussion with the employee. Such discussion is an extremely valuable part of the employee development process and should cause us to reflect carefully on our trends toward automation. It should be apparent at this point that this entire process of generating a personal learning matrix, conducting a self-assessment, prioritizing development needs and creating an individual development plan lends itself very well to automation. There are several human resource or employee development software packages available today to do exactly that. I am not saying that such software should not be used, but that it should be used with caution. When the automated approach is available, and everyone is feeling way too busy with their other projects, it becomes easy to allow the employee to fill in a few electronic blanks, 'rubber stamp' the supervisor's mechanized approval and move on to other projects. Engineers and scientists may be especially prone to this technological shortcut.

Returning now to the one-on-one discussion between a supervisor and an employee we are positioned to make some valuable discoveries.

7.1.1 Case 1

Stephen has been an engineering team leader in the same department for six years. Recently Michael transferred from another part of the company, and is now on Stephen's team. Michael has only been with the company for two years and, as Stephen knows, he joined immediately after completing his engineering degree. Stephen is concerned because in Michael's self-assessment he has claimed a high proficiency with a particular, difficult manufacturing process important to the machining of their team's product. He is sure that, having recently finished college, Michael has misunderstood the intended subject and proficiency expectations and what would be expected of him at the proficiency he claims. Rather than just downgrading the proficiency Michael claims, Stephen decides to talk with him about it, with the intent of explaining to Michael why it will be important to reassess this proficiency claim. In the course of the conversation Stephen learns that Michael is in fact quite an expert in this subject – his father owns a high technology machine shop, specializing in exactly this type of machining, and Michael worked in this shop throughout college, in several cases personally completing critical process development projects very relevant to the work of Stephen's team. Not only has Stephen learned that Michael's self-assessment was accurate, but in the process he discovered a tremendous resource within his own team for mentoring other team members.

7.1.2 Case 2

Valerie is leading a new team assembled with experienced people from throughout the company to work on an important project. John is a chemist assigned to her team to lead the lubricant studies associated with the project. However, in reviewing the self-assessment John has just completed, Valerie notes with alarm that he professes very limited proficiencies in lubricant analysis. It is clear from John's self-assessment that he possesses strong proficiencies in a wide variety of areas, but there is a glaring hole in virtually the exact subject area for which John has been assigned to this project! In talking with John, Valerie learns that no one in his old department talked with him before recommending him for the new assignment. While he was pleased to be chosen he was never told what his new responsibilities would entail. The management of his former department merely assumed that he had considerable personal experience in lubricant analysis, when in fact his experience had been in managing a small, easily overlooked sub-contract to an outside firm that did the lubricant analysis. John is a willing and able learner, and still has several professional contacts in this area. Valerie has decided that John can work with his contacts, both to complete the lubricant analysis that will soon be needed and learn more about the analysis processes, as more advanced work will be required later in the project.

7.1.3 Case 3

Seth recently joined the company after completing his doctorate at a prestigious university. He is anxious to demonstrate his capabilities to his new employer, and is delighted that he was soon asked to complete a self-assessment of his proficiencies. Most of the subjects the company has listed as relevant to his position are in areas he recently covered in his courses – in all of which he did very well. He confidently claimed high proficiencies in several key subject areas and was sure he would quickly be given the responsibilities to lead important aspects of the project to which he is to contribute. Ian is Seth's supervisor. While he has nowhere near the formal education of Seth he has over twenty years of experience with the product, and had to suppress a chuckle when he reviewed Seth's self-assessment. There was nothing to be done but to sit down with Seth and go through the subjects one by one. He talked with Seth about each subject and how it was applied to their product; he asked Seth many questions about how he would address particular problems – often far more complicated than could be simulated in a laboratory. While this was a humbling experience for Seth he was thankful that it occurred in the office of this experienced and helpful engineer, and not in the midst of his project work, where his fellow team members would have been forced to correct him. Ian and Seth concluded their meeting with the first of what would become frequent sessions addressing the continued development and growing application of Seth's strong proficiency base.

In two of the three cases just described the disagreement between the supervisor's assessment and an employee's self-assessment were resolved with the employee having in fact made an accurate judgment. This will certainly most often be the case when the expectations versus proficiency level have been clearly defined. In all three cases the information gained through the individual discussions resulted in actions that were very beneficial to the project teams. This serves to underscore the important role played by the supervisor, not in conducting the assessment, but in reviewing and clarifying the assessment made by the employee.

7.2 Supervisor review of priorities

Once the employee and supervisor agree on the employee's current proficiencies priorities must be set for the employee's continued development. This topic was previously discussed from the employee's perspective in Chapter 6, Sections 6.3 and 6.4. In that discussion the following criteria were identified for deciding the priority to be placed on each learning need:

- current project needs;
- anticipated, near-future project needs;
- employee's short-term and long-term career development interests.

The reader is referred back to that discussion and the example of Janet, the team leader for development of an engine lubrication system, to review how employee development priorities are determined.

Returning to the discussion from the perspective of the supervisor I would emphasize only that the supervisor must keep in mind not only immediate project needs but the employee's career interests and resulting long-range development needs as well. Failure to maintain this balance may result not only in stifling the individual employee's career, but in failing to deliver on immediate project needs as well. This will be further demonstrated in the next section.

In a very real sense the company is entrusting responsibility to the supervisor for the development and maintenance of human resources critical to its success. The retention, satisfaction and growth of each employee entrusted to a supervisor should certainly be one of the measures by which a supervisor's performance is assessed. The supervisor must be equipped to provide accurate and helpful career guidance to each employee, and ensure that development needs relevant to that employee's career goals are being addressed.

7.3 Supervisor role in development plan

The following scenario will provide a practical demonstration of the roles of the supervisor in the employees' development plans.

Michael manages the structural analysis group for a leading manufacturer of steam turbines. Brian has worked for Michael for several years. Two years ago Brian developed a very nice technique for estimating the fatigue life of turbine blades and correlating it to field test results under various duty cycles. He used this technique last year to recommend design changes that allowed a 30 percent up-rate of a popular product with no loss in durability. This successful project resulted in very favorable attention from several of the company's officers. Another family of turbines produced by the company is currently seeing premature turbine blade failures. The problem is becoming quite costly and if it is not soon corrected it will be damaging to the company's reputation. Not surprisingly, Michael has been assigned responsibility to lead in solving this problem.

Brian and Michael have recently discussed Brian's career interests. Brian is a fatigue specialist who wants to continue to gain expertise over a variety of fatigue problems. He really enjoys the challenge of developing new solution techniques. They talked about Brian's interest in characterizing thermal fatigue in the steam nozzles. Brian has some interesting ideas that will streamline this design process. This will be quite beneficial next time a new nozzle is required, but any such need will be at least a year away.

The scenario just described represents a dilemma commonly faced by engineering managers, and one that is more often than not handled poorly. To many engineering managers the solution seems all too obvious. Yes, Brian would like to do something new – wouldn't we all? But he's being paid by the company to contribute what is best for the company, and right now solving this blade failure problem on a production unit is critical. Michael will talk with Brian, explain the situation and assign him this critical project, of course reminding him of why it's in his best career interests to demonstrate another successful solution to the company leadership.

Michael has chosen the obvious solution. Because Brian is so familiar with this analysis technique he will be able to solve the problem more quickly and effectively than anyone else on Michael's staff. As far as most engineering managers are concerned this is the only solution that makes sense and it often seems to work quite well. But Michael has missed several important points and whether he realizes it or not he is damaging the effectiveness of his organization, and in the long run costing his company money every time he makes this seemingly obvious project assignment decision. The points he is missing are:

1 Every time such a decision is made the morale of the organization is negatively impacted. While the results occur slowly and would be almost impossible to measure, he is affecting the productivity, capability and employee retention of his organization.

2 Employee development is not some discrete event one periodically sends the workforce off to, but is an integral part of the day-to-day project assignments given to each employee.

While Michael thinks he has properly fulfilled his responsibilities as a supervisor he may have just made the assignment that helps Brian with a decision of his own. Perhaps this isn't the first time that Brian has been assigned to a project based on his current expertise. Brian is an extremely capable engineer and he thrives on challenge. Every assignment he's given based on what he's already good at is more boring and less challenging than the last. The job market is really strong and the company's biggest competitor is ready to offer Brian an attractive role. Perhaps this is the time Brian decides he's had enough, and instead of a successfully completed project, Michael receives Brian's resignation letter. Oh, and one more thing . . . that particular fatigue analysis technique Brian perfected was a little more complex than anyone realized and if Brian had a serious fault it was his unwillingness to keep written documentation of his work.

This may seem like a worst-case scenario, but in fact it is all too real and costly a problem for many engineering organizations. It becomes especially costly when a talented and experienced engineer carries his intellectual capital directly to the competition.

What alternative did Michael have? Brian was not the only engineer in his department. Jeffrey joined the department six months ago after completing his masters degree. He has little experience, but it's already clear that he learns quickly and just about any project he is assigned will benefit his development. Michael assigns Jeffrey to the project, but before top management has time to question his sanity he assigns Brian as Jeffrey's mentor on the project. Jeffrey can do much of the work under Brian's guidance, and nearly half of Brian's time can be freed up to begin working on that nozzle thermal loading problem. Yes, the blade fatigue solution requires slightly more time (but far less than if Brian had left the company!). And certainly something else had to be dropped in order to assign Jeffrey to this problem and allow Brian to begin working on the nozzle loading problem, which had been 'below the line.' In the long run this will be a small price to pay for employee development and employee satisfaction. On the plus side, let's quickly review what's been gained:

1 Brian has been given a project that's personally challenging and is nicely aligned with his career development needs.
2 Brian's skills in an area of critical importance to the company have been passed along to a sharp, new engineer.
3 The blade fatigue problem did in fact get solved and the increased time required was probably quite minimal.

4 Next time a nozzle needs to be designed, a new analysis technique will be available to significantly improve the process.
5 The morale of both Brian and Jeffrey is quite high, and positively impacts everyone else in the department.

Some of these results will be difficult to measure, while others will be quite apparent. All of them are very real, and valuable to the organization.

The scenario with which Michael is faced is summarized in general terms in Figure 7.1. In the figure the possible sequencing of events flows from left to right. The critical project is identified, and the supervisor makes assignments. The upper portion of the figure (solid boxes) represents the seemingly obvious, or pragmatic, approach to project assignments. The lower portion (dashed boxes) represents the alternative approach described in this example, in which the focus is simultaneously placed on employee development. While the emphasis on development may result in a slight delay in completion of a critical project, it simultaneously better prepares the organization with new learning for the next project, and reduces the risks and costs associated with employee dissatisfaction.

The supervisor's role in employee development should be quite clear from this example and the summary given in Figure 7.1. It begins from the recognition that employee development is part of the job – not a separate activity. **Project assignments should be made by building on each employee's strengths while simultaneously stretching them to gain proficiency in areas that complement their career goals. Each employee's current strengths should be utilized as a resource for the development of others**. These simple guidelines should form the backbone of each employee's development plan. Formal training, and other employee development approaches, should supplement the role of on-the-job learning.

7.4 Corporate-wide support

In addressing the role of the supervisor in employee development it is important to consider not only the direct supervisor, but the company's overall management and the resulting framework for development. It will be the role of management to communicate to each supervisor the value of their roles in employee development. It will be the role of management to ensure that tools are in place to support the learning process. The learning matrix database has already been discussed as one such tool. Further tools might include communication and education for supervisors and individuals to ensure that everyone understands their roles and the company's vision regarding employee development. For example, the vision being presented here includes the following previously developed points:

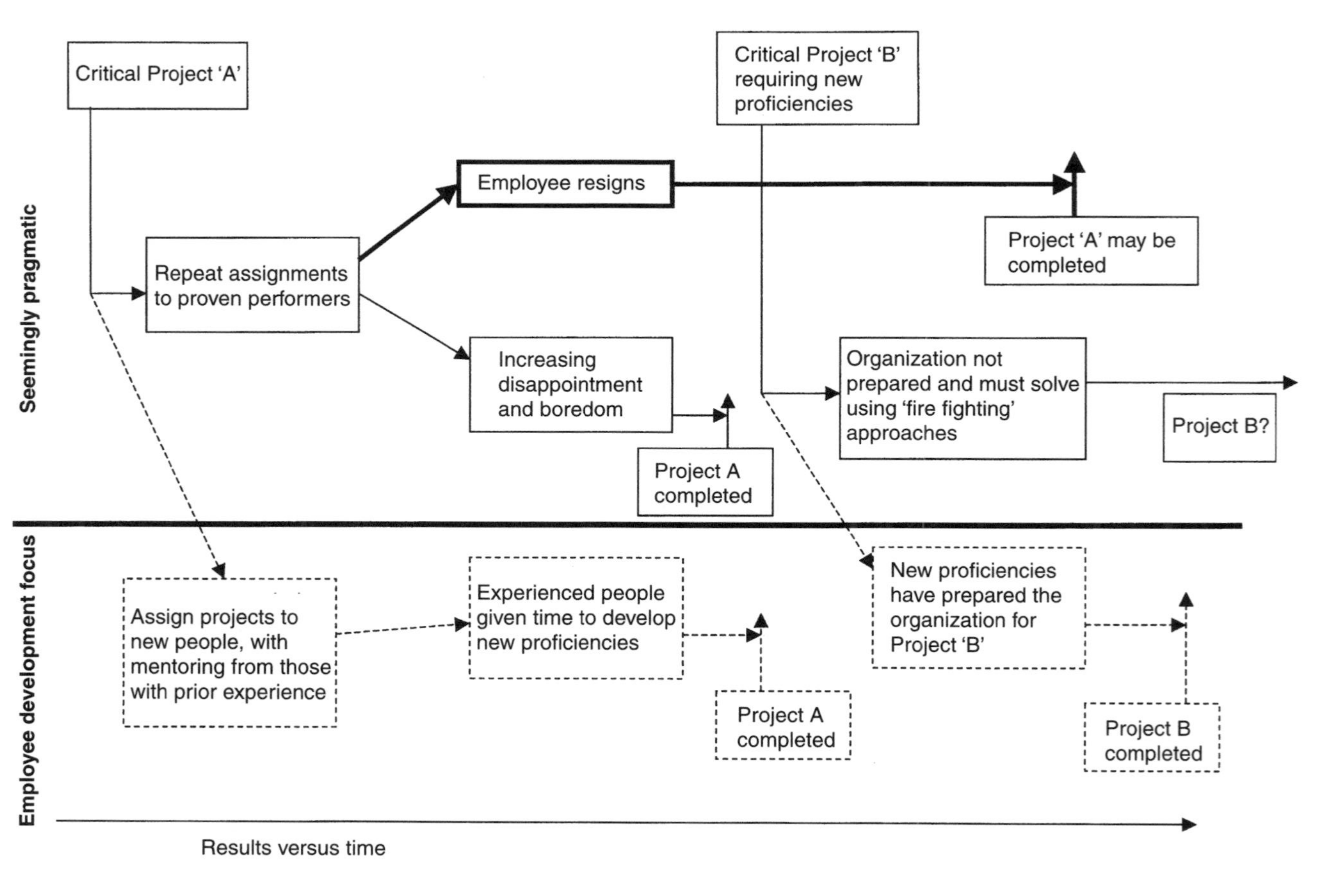

Figure 7.1 The importance of viewing skill development as part of the work

- employees hold primary responsibility for their own learning;
- immediate project needs and employee career interests must both be addressed;
- new learning is an integral part of each employee's project assignments.

Finally, the company should provide a common format to ensure that development discussions are taking place, and that these discussions are aligned with the company's goals. Based on the vision being presented in this book, a simple development planning form documenting the following information will be very effective:

- a brief summary of the employee's short-term and long-term career goals;
- a tabulation of the employee's current proficiencies and prioritized development needs. This may be similar to the example given in Chapter 6, Table 6.3;
- a documented development plan emphasizing work assignments and, supplemented with training programs or other development activities.

This simple form should be updated through one-on-one discussions between the employee and supervisor on a regular basis. It should be emphasized that the simple system described here, if carefully implemented, not only provides a very effective employee development system, but not surprisingly, also fulfills the ISO certification requirements pertaining to employee development. This is a much more manageable (and effective) system than the spreadsheet training tracking systems most often employed to address ISO certification requirements.

As was mentioned earlier in this chapter, proficiency tracking and documenting the development plan can be done electronically, and doing so may hold several advantages. As will be discussed later the resulting database can be cut in any of a variety of ways. Examples include assessing technical strengths throughout the company, to plan for and staff new projects, or to track employee learning over time. Again it must be emphasized that special care must be taken to ensure that these powerful and time-saving tools do not drive the element of personal discussion out of the process.

7.5 Summary

- While the employees should individually assess their own proficiencies, it is important that the supervisor carefully review the employees' assessments.
- Discussions of any discrepancies between the employee and supervisor assessments will result in one of the following:

  - supervisor discovers employee proficiencies not previously known;
  - employees receive development they may not otherwise have received;
  - employee gains more accurate knowledge of company expectations.

- The supervisor plays the key determining role in prioritizing each employee's development plans.
- It is absolutely critical that the supervisor grasp the importance of treating employee development not as something separate from the work but as a fundamental component of each project. This is necessary in order to:
  - build and maintain employee satisfaction;
  - ensure department capabilities are in place in advance of when they are needed;
  - provide succession plans in case of employee retirements, transfers, or resignations.
- The employee development process can be simply and effectively tracked using a form containing the following information:
  - summary of employee's short-term and long-term career goals;
  - tabulation of employee proficiencies and prioritized development needs;
  - development plan emphasizing on-the-job learning supplemented in some cases with formal training programs.

PART 2

A FEW FURTHER APPLICATIONS

The first part of this book presented a model for encouraging and supporting employee development in the workplace. The model emphasized developing advanced proficiencies in technical subjects and encouraged a focus on on-the-job learning. At this point a major objective of this book has been satisfied. The material presented up to this point completes the picture of the advanced learning model and its implementation.

The remaining chapters cover related topics and demonstrate a few further workplace tools that can be supported by this learning model. Chapter 8 discusses the use of rotational programs as a means of further encouraging project-based learning. Rotational programs are often thought of as a rapid cycle through a variety of organizations, as is often done with new employees to familiarize them with a large engineering organization. In this case the emphasis will be on longer-term assignments, but with the use of pre-planned transfers designed to provide the employee with a desired base of experiences.

In Chapter 9 the topic of organizational assessment is taken up. This topic was previously mentioned in earlier chapters as a resulting benefit of the advanced learning model. Here the topic will receive more detailed attention, and several examples will be used to demonstrate its effective use.

Finally, in Chapter 10 the subject of evaluating particular training and learning interventions will be addressed. It is often an objective of the senior managers to determine whether money spent on a particular training program was worthwhile. They would like to know whether the employees who participated in the program learned anything, and more importantly (but more difficult to determine), whether their increased proficiency made them more effective employees – more effective enough to have justified the training investment. It will be shown that with relatively little additional effort the learning model presented in this book can be used to gain valuable information regarding the effectiveness of not only formal training programs, but the on-the-job experiences and the various other learning mechanisms described earlier in Chapter 4.

Chapter 8

Rotational programs

8.1 Introduction

Much of the emphasis of this book has been on the techniques that can be used to develop advanced proficiencies. Several of these techniques were discussed in Chapter 4, and it was shown that such proficiency development occurs not so much in the classroom but primarily through on-the-job experience. As employee development recommendations are created, the emphasis is placed on learning through doing – often under the guidance of a mentor or other expert in the discipline being developed. It has been argued that a great deal of learning can occur through careful selection of project assignments and explicit attention to assigning projects that build from the employee's current proficiencies, while allowing the employee to stretch into new realms, and thus develop new proficiencies. This process can be furthered through the use of planned rotational programs specifically designed to meet both individual and organizational needs for increased breadth of understanding. The focus of this chapter is on such programs. The basic goals of long-term rotational programs (as opposed to relatively short-term orientation programs) are discussed. A discussion of the processes involved in developing the program that will best meet a particular company's overall learning goals will then be presented. It will be shown that this process follows directly from the learning matrices developed earlier in the book. Finally, a specific case study will be presented to demonstrate the concepts, and the experience gained through developing and implementing the program will be shared.

8.2 Description of the concept

Perhaps the best place from which to begin a discussion of planned rotational programs is by emphasizing what they are not. Many companies, especially larger companies with several divisions, or complicated organizational structures, have developed formal orientation programs for new employees. Lasting anywhere from a few months to a couple of years, such programs consist of an array of brief, familiarization assignments and are designed to acclimate employees to the organization, and in some cases to aid in determining where the employees might best contribute in their permanent roles to follow. This is NOT the type of program to be discussed here.

As opposed to a relatively brief orientation program, a planned rotational program consists of a sequence of three or four long-term roles specifically planned to provide the breadth of experience desired in future technical leaders. Each assignment represents full engagement in one or more multi-faceted projects, and is not unlike the roles that might be taken on by an individual periodically embarking upon transfers based on personal interests or organizational needs. The only real difference is that the sequence of assignments, and agreed upon schedule, has been determined at the outset based on pre-determined objectives for the individual's development. The program objectives may include broad-based technical learning, as well as the development of expertise in product application and customer requirements, manufacturing processes and possibly managerial or project leadership proficiency.

The distinctions between rotational programs, orientation programs and traditional transfers are further summarized in Table 8.1. As was stated at the outset, the rotational assignments are of sufficient length to more closely resemble a full-time assignment to a particular project. The timescale is compressed somewhat, but is designed to ensure that

Table 8.1 Summary of comparison between orientation programs, planned rotational programs and traditional transfers

	Orientation programs	Planned rotational programs	Traditional transfers
Length of time in assignment	2 weeks to 6 months	1 to 2 years	2½ to 5 years
Level of project responsibility	Low	High	High
Depth of learning	Low	Medium	Medium to high
Breadth of learning	High	High	Typically low

the individual can hold significant project responsibility and gain a reasonable depth of learning. The result of a well-designed rotational program is a pipeline of engineers and scientists possessing a broad base of solid technical experience. Traditional organizational transfers can theoretically provide similar results, but without close attention and guidance they will seldom provide the breadth of knowledge most companies would like to see in their future technical leaders.

Finally, an appropriate balance between organizational needs and individual career interests must always be maintained. This might best be done by creating a general framework that addresses the organization's interests regarding the areas of expertise to be developed. From this general framework, the employee may be allowed to make personal choices from an array of options. Of course the ability of an organization to offer such flexibility will be quite dependent on their needs at any given time. It is certainly in the organization's best interest to offer as much flexibility as possible, since employee frustration or dissatisfaction goes a long way toward undermining the value of the long-term investment in employee development. While the employee's contributions will have been every bit as real and valuable as those from any other employee who then leaves the company, the resignation of an employee having completed perhaps 80 percent of a planned rotational program will be especially difficult to accept.

8.3 Developing the rotational program

The learning matrices developed as described in Chapters 3 and 4 form a tremendous starting point from which a planned rotational program can be created. Chapter 3 described chartering a team of senior technical people within a particular discipline to identify the subjects and proficiencies needed by engineers working in that discipline as they progress in their careers. In Chapter 4 these same teams were called upon to look beyond classroom learning and identify recommended approaches through which engineers could gain increased proficiency – these approaches were generally found to be very closely tied to their project assignments. This effort in fact lays a strong foundation from which the rotational program proceeds.

The extension that needs to be made from the previous work is that of the breadth of subjects covered. The learning matrices were developed within particular disciplines, and focused to a great extent on continued growth within that discipline. In other words, there was a tacit assumption that the engineer would spend an entire career within the particular learning matrix. While this is true for some people, in most engineering organizations this represents a small fraction of the engineering workforce. Many more engineers have wider-ranging interests, often with career goals of ultimately leading multi-faceted projects requiring the application and management of expertise from a variety of disciplines.

The objective of the planned rotational program is to develop the breadth of proficiencies required by the future project leaders or chief engineers. With this objective in mind the next task is to determine what a new learning matrix would look like, combining a subset of the subjects from several matrices, but never achieving the depth of proficiency in any one specialized subject area. In other words, the thinking shifts from depth of knowledge to breadth of knowledge. The list of subjects is drawn from several disciplines and the proficiencies required at any given organizational level are revised. This revised learning matrix is drawn up based on success criteria for those in project leadership positions. The team involved in drawing it up may be made up of technical experts from various disciplines, but in most cases should instead be made up of successful project leaders.

Because the subjects placed in this new learning matrix should all come directly from the existing matrix, the work required in order to decide how best to gain those proficiencies will also already have been done. This provides an excellent base from which to create a rotational program. The kinds of projects and employee development opportunities have already been identified, and the only remaining tasks are to determine the order of rotational assignments, the length of each assignment (based on estimates of how long it would take to develop the particular proficiencies) and the basic ground rules of the program.

In determining the order in which assignments should be completed two questions must be answered. The first of these is to what extent the proficiencies developed in one assignment will be required for success in another. This may be determined by randomly laying out the particular disciplines or assignment areas as shown in Figure 8.1. If one selects any two assignments an arrow should be drawn from the area in which proficiencies should be developed first, to the area that can be developed second. In some cases the proficiencies from either area may not be

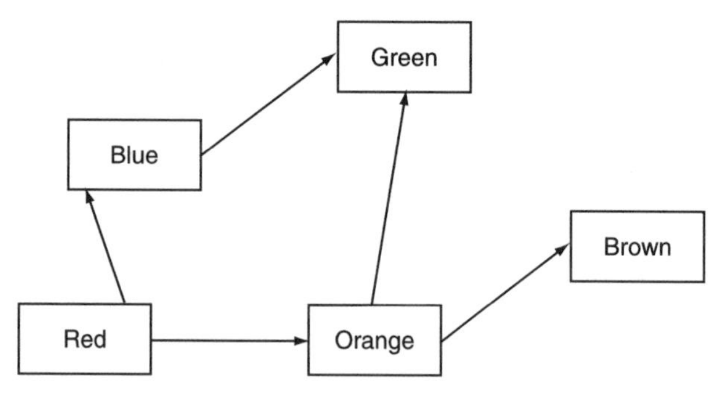

Figure 8.1 Determining the order of rotation

required for the other, and no arrow need be drawn. This would imply that assignments in these areas could be completed in either order. This in fact may add flexibility to the program and ease the flow of engineers between assignments, as will be shown in the next section. Once the placement of all the arrows has been determined the basic flow of the rotational program has been defined.

In Figure 8.1 the assignments have arbitrarily been assigned a color for identification. Based on the arrows that have been identified, all engineers entering this rotational sequence would need to begin with the red assignment, as the proficiencies developed while in that assignment will be necessary for all others. From the red assignment employees could proceed to either the orange or the blue. This provides some of the flexibility described earlier, as the number of engineers transferred to each organization can be determined based on current project needs at any given time. From either the blue or the orange assignment engineers could then be transferred to the green. Here again there is some flexibility in that a portion of the engineers could move from the orange to the brown assignment, instead of the green.

A second question must be addressed at this time – do all engineers need to rotate through all of the assignments, or are some of the assignments similar enough that an engineer passes through one but not the other? A modification of Figure 8.1 is given as Figure 8.2. If the arrows shown in Figure 8.1 are now taken to indicate the actual rotational paths, some engineers take assignments in the blue area and others in the orange. Most engineers then transfer into the green organization while a few transfer into the brown. Referring now to Figure 8.2 it has been decided that engineers need experience in both the blue and the orange areas before proceeding to the green or brown, and that when they complete their assignments in either the green or brown they proceed to the other area.

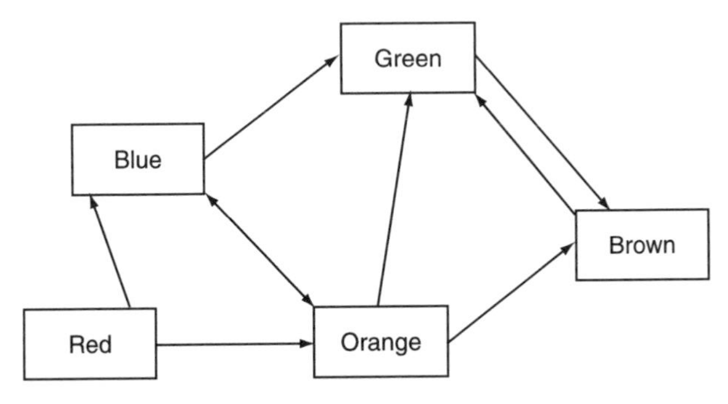

Figure 8.2 Modified rotation to ensure that all engineers experience all areas

8.4 Cummins case study

It will now be instructive to take the concepts discussed in the preceding sections and demonstrate their application in a specific company. This example is taken from the author's experience at Cummins Engine Company, Inc.

The rotational program at Cummins, referred to as the **Engineering Development Program** (EDP), was developed to address two issues identified by the engineering leadership. First, it was observed that several of the company's best engineering project leaders would soon be retiring and that there were very few engineers believed to be technically prepared to move in to these roles. The internal combustion engines produced by Cummins are very complex products, calling upon expertise in a wide variety of engineering disciplines. Few engineers develop the breadth of experience required to lead the integration of these disciplines and thus coordinate the development of a new product.

The second issue identified by the engineering leadership concerned the company's hiring patterns. The vast majority of new engineers hired were being placed in product development roles. Much of the work of the engineers in these roles is driven by very tight time schedules. There is little time to reflect on root causes, or even to learn the fundamental concepts and engineering tools specific to the product.

Based on the two issues just described, the goals of the Engineering Development Program were to create a pipeline of well-grounded engineers flowing into product development roles from within the company, and to create a development path that would ensure that engineers were gaining the experiences needed in order to take on future project leadership roles.

The effort of assembling the Engineering Development Program began with a team of engineering leaders discussing the technical subjects and proficiencies in each subject needed by a Chief Engineer. One of the problems immediately identified was the fact that the core of internal combustion engine development involves two quite different disciplines from within the traditional realm of mechanical engineering. One is that of design, and the static and dynamic analysis required in developing reliable and durable engine components and systems. The other is that of the thermal sciences, which includes proficiencies in engine performance, fuel economy, emissions, and cooling and lubrication systems. Left to their own devices, very few engineers will gain experience in both of these broad discipline areas. Engineers will typically have a preference for one or the other and their career histories will reflect a series of assignments most often remaining on one side or the other. As a Chief Engineer some degree of background in both the mechanical development and the performance development of the engine is desired. This observation, and the desire to ensure that engineers become well grounded in the fundamental sciences and engineering tools, provided the basis for determining the initial assignment options in the Cummins Engineering Development Program.

Two entry points were identified, both in central, technical support groups that conduct advanced analysis and experimental work in support of the various product development groups. One of these groups was an organization focused on thermal sciences, referred to as 'Performance and Combustion.' The other was a mechanical development group known as 'Design and Analysis.' Both of these groups were staffed with experienced engineers who possessed great depths of expertise in their specific fields. Prior to implementing the Engineering Development Program these groups hired very few young engineers directly from college. In order to address the need for engineers to gain experience in both the mechanical and thermal sciences, it was determined that new engineers would join the company in one of these groups, and upon completion of their initial assignment, rotate to the other group. Learning matrices for each of these assignments were created as subsets of the subjects previously identified for engineers in these organizations. Careful review of the desired subjects and proficiency levels led to the conclusion that each of these assignments should be 18 months in length.

The initial portion of the Cummins program is thus summarized in Figure 8.3. Engineers enter at one of the two locations shown, spend 18 months in each of the two locations, and then transfer to the next phase of the program. Again, asking what subjects and proficiency levels are needed by a Chief Engineer resulted in identifying the next assignments as customer-focused roles. Three different types of roles were identified within the Cummins organization. These were Service Engineering, Application Engineering, and Manufacturing and Quality Engineering. Service Engineering roles involve direct interaction with the customers of the products. Application Engineering roles involve working with vehicle manufacturers installing the engines in their products. Finally, Manufacturing Engineers are those involved in producing the product, and so

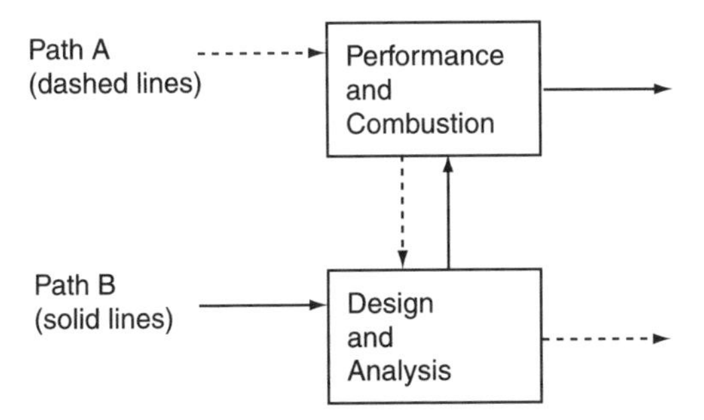

Figure 8.3 First phase assignments in Cummins' Engineering Development Program

are seen as the direct customers of the product development process. It was determined that the engineer would gain an excellent foundation through experience in any one of these areas. The second phase of the program thus consist of a twelve-month assignment in one of these three areas. The specific area would be decided by a combination of employee interests and current project needs at the time of the transfer. Adding the second phase, the program appears as shown in Figure 8.4.

An interesting finding as the program was implemented was that, while the engineering leadership saw an equivalency among these three types of roles, the participants in the program viewed manufacturing experience quite differently from either service or applications experience. Several of the first engineers in the program asked whether they could gain experience in two assignments during this phase. In keeping with the emphasis on employee development, as opposed to orientation, it was not believed that the assignments could be shortened from the allotted twelve months, so the engineers were given the option of adding an extra year to the program, and taking two roles during this phase. The stipulation was that one of the two roles then had to be in manufacturing, with the other in either service or application engineering. Since that time approximately 50 percent of program participants have opted for a second phase-two assignment.

Recall that one of the objectives of the Cummins program was to create a pipeline of experienced engineers flowing into product development roles. It was therefore decided to complete the program with an assignment in a product development role; it was further decided to

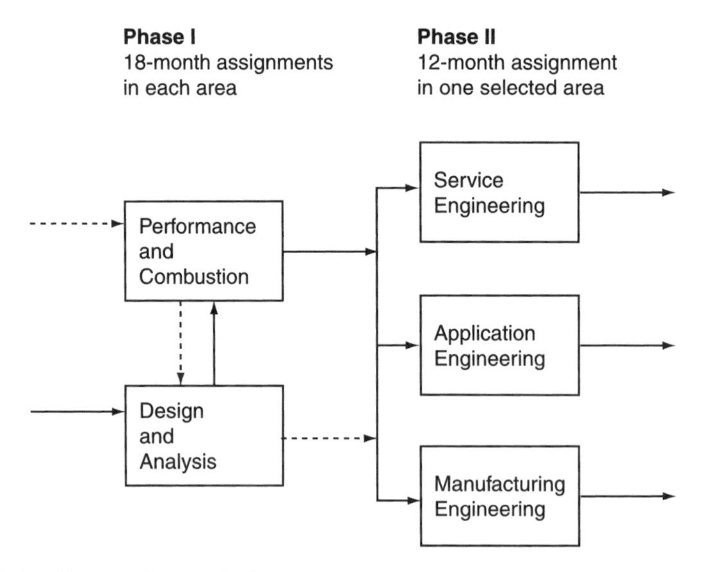

Figure 8.4 First and second phase assignments in Cummins' Engineering Development Program

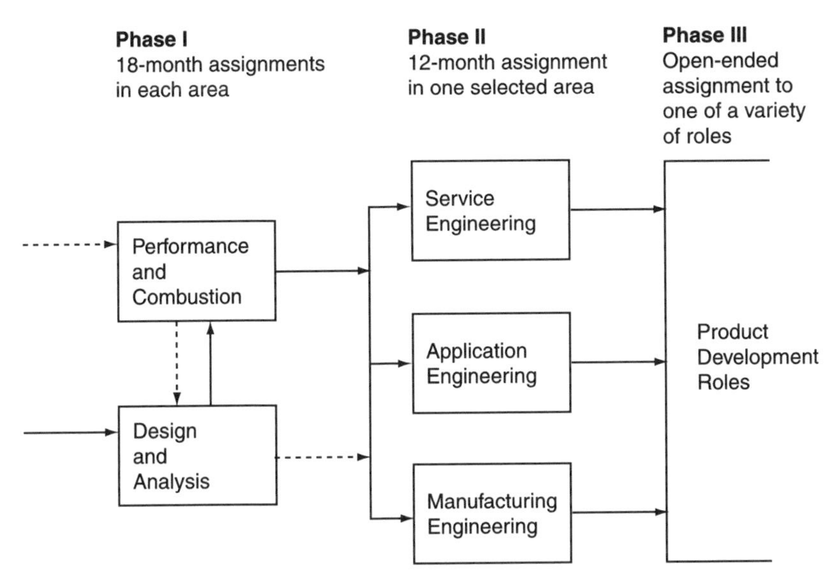

Figure 8.5 The complete Cummins' Engineering Development Program

make this an open-ended assignment. At the conclusion of the program the employee remains in the product development assignment until such a time as she or he chooses to transfer to another role, or the company arranges for a new assignment. The product development assignments, shown as a single, open-ended box in Figure 8.5, actually represent a wide variety of roles in either advanced or current product development, over a wide range of products. This variety allows a great deal of flexibility in selecting the final role for each Engineering Development Program participant. This provides a great deal of opportunity to match company needs at a given point in time to individual employee interests.

8.5 Summary

- Planned rotational programs can provide an excellent mechanism to further support on-the-job learning.
- Planned rotational programs differ from orientation programs, both in terms of the length of assignments and the level of proficiency developed in each area. As such they must be viewed as an integral part of an employee's career rather than something that must be put up with prior to engaging in 'real' assignments.
- The rotational program must be developed based on identification of the long range proficiencies required of future technical leaders, and the work assignments best suited to developing those proficiencies.

- The order in which employees progress through the rotational program is determined by the requisite proficiencies for each assignment.
- Where two or more assignments do not depend upon one another for proficiency development, the rotational program gains flexibility; employees can be assigned based on project needs at the particular point in time.
- Where two or more possible assignments can be utilized to achieve similar proficiencies options can be offered that give the employee further flexibility to address personal career interests.

Chapter 9

Organizational assessment

9.1 Introduction

Throughout this book the subject of addressing the overall learning needs of a technical organization has been mentioned periodically. The tools are now in place to take a more formal look at this subject. This chapter begins with a summary of the needs an organization faces to assess its proficiency across an array of technical subjects. Once again it will be shown that the tools developed in this book provide an excellent framework from which organizational assessment can be conducted. The chapter will show how the data obtained from the learning matrix database and employee self-assessments are used to address the overall needs of the organization.

9.2 The needs of the organization

Among the most often discussed, and least often satisfactorily addressed, needs of any company are those pertaining to human resource planning. The following paragraphs describe several examples of these challenges; most will be quite familiar to managers and human resources professionals.

9.2.1 Department proficiency assessment

Perhaps the most obvious, or at least the most ongoing, need is to rapidly and accurately assess the current strengths and development needs of a department or project. While this might seem quite self-evident, it is remarkable how few managers can produce an accurate picture of their own department. When requested, what will more typically be provided is a partial assessment. The manager will generally be able to identify a person who is especially strong in a particular subject – one that has recently been successfully demonstrated on a high visibility project. But she or he may be quite unaware that the same person possesses strong proficiency in another subject of great interest to the company, but which is currently not being tapped. Similarly, the manager may be able to point to a person having a particular development need – unfortunately, more often than not, only because that person recently stumbled on a high visibility project. But once again, it is much less likely that other development needs or career interests will be readily identified. Even the strengths and development needs identified by the manager may be much less accurate than supposed. This is due to such factors as the manager not being fully familiar with the current state-of-the-art in a given subject area; the employee successfully laying claim to a high proficiency based on innovative use of smoke and mirrors; or inaccurate comments or reports from others in the department, or elsewhere in the organization. A manager will be far more effective in maintaining and advancing the strengths of the department if an accurate, annually updated department proficiency assessment is available.

9.2.2 Succession planning

Following directly from the preceding paragraph is the often discussed, but seldom well-implemented, succession plan. It is always in both the manager's and the company's best interest to have a plan in place that allows people to make the career moves they want, and addresses upcoming retirements and the possibility of unexpected attrition. Combining an accurate, department-wide (or by extension, company-wide) proficiency assessment with the individual development plans discussed in Chapter 7, provides an information base from which succession planning can be solidly implemented. Specific cases of individual retirements and desired career moves can be planned for in advance, as the manager can

- identify the subject areas that will be lost to the department;
- identify people especially interested in developing proficiency in those subjects;
- utilize project assignments, mentoring and other learning techniques to develop the needed proficiencies.

This process may not always immediately and completely fill the gaps in the department's capabilities, but it sets the process moving in the right direction.

The more difficult situation is that of sudden and unplanned attrition. While this may be reduced by taking the steps discussed in Chapter 7, it remains a fact of life throughout the employment world. The manager is best served by continuing to keep an eye on the department's strengths and development needs, and as much as possible ensure that more than one person possesses reasonably strong proficiencies in any subject area of ongoing importance to the department. The definition of 'reasonably strong' will vary widely from one company, or even one department, to another.

9.2.3 New project staffing

In many companies the need periodically arises to rapidly staff a new project. This may be done for a variety of reasons including the development of a new product, or addressing a customer or manufacturing problem. While such needs are often addressed by hiring from outside of the company, the focus of this paragraph is on projects that are staffed by reassigning people from within the company. Such reassignments create challenges for both the managers trying to staff the new projects, as well as the managers who lose a person to the project. The situation for those losing a person is not unlike the case of sudden attrition, described in the previous paragraph, and must be handled similarly. The question taken up in this paragraph is how best to rapidly and accurately identify people from throughout the company (or the available locations within the company) to staff the new project. Without the company-wide proficiency assessment data, the selection of staff is made based on friendships and personal knowledge, the recommendations of colleagues and other, still less scientific and more problematic, grounds. At the very least, such methods result in selections made after considering only a portion of the proficiency base. It often results further in a less competent project team than might otherwise have been possible. The availability of an accurate snapshot of the proficiency base and development needs from around the company overcomes these problems, and ensures that the new project team will possess the best of available talent in each critical area. It may also do less damage to other parts of the company; if two people from different departments possess similar capabilities, the one selected can be chosen from the department having greater depth in this subject area.

9.2.4 Finding the talent and diversity

In the preceding paragraph it was stated that a company-wide snapshot of the technical proficiency base allows one to better ensure that the best selections are made when filling key roles or staffing new projects. It was also pointed out that in the absence of such data, positions are often filled based on friendships and the internal networks that develop within a company – word of mouth recommendations among colleagues. It is often legitimately argued that such an approach to staffing – especially when looking for people to fill key, high visibility positions – unwittingly becomes a contributor to the 'glass ceiling' that keeps women and ethnic minorities from advancing beyond mid-level positions. Although unintentional, this problem is very real and a very direct result of the limited data available to the manager when staffing selections are made. An example will demonstrate the problem.

Michael is a Chief Engineer who has just been given the responsibility to lead a team in developing a ride control system for a luxury automobile. The automobile manufacturer is a new customer for Michael's company, and this particular car is their flagship model. Not surprisingly, this project has the attention of the company president and his Vice Presidents of Marketing and Engineering; they have already informed Michael that they would like bi-weekly updates from him and his team throughout the development process. Michael realizes that the success or failure of this project will rest on the caliber of the team he is able to assemble, but this has him quite concerned. His own background is almost entirely in design and manufacturing, and it is already apparent that among the most critical roles will be that of an expert in controls algorithms, and someone with expertise in sensors and actuators.

Fortunately for Michael he remembers meeting one of the company's sharp young engineers at his health club last month. The young man impressed him a great deal – not only on the squash court, but later, in the locker room, when he talked with Michael about how much he enjoyed his work in controls algorithms development. He would be a natural and, judging by his personality, he would be a natural leader. Then as luck would have it, Charles, a Chief Designer with whom Michael has worked for many years, approached him after church, said that he heard Michael was looking for an expert in sensor and actuator design, and introduced him to his young friend who just joined their church. The introduction went very well, and Michael concluded with delight that he had just filled two important positions on his project.

To almost anyone having experience in a large company, the example just cited will sound quite familiar. In the absence of solid data upon which one can make such selections, most are made in ways very similar to those just described. In most cases the approach is quite effective and the selected people make significant advances in their careers. However, there are at least two problems with this approach. The first stems from the general observation that Michael in this example has no objective measure of how well qualified the selected people are relative to others in his company, and no idea whether these were the most strategic choices based on the wider needs of the company. The second is a specific problem relative to enhancing the diversity of the company in its future leadership ranks. Because of the circumstances under which each person was identified the pool of potential candidates was narrowed to those having similar interests and cultural backgrounds to Michael and the current leading people in his company. This selection approach, if used widely within any company, results in a remarkably similar-looking leadership profile, and effectively excludes people having dissimilar interests and backgrounds from leadership roles.

Once again, having a company-wide database of the current and developing expertise ensures that all qualified candidates can be considered for important new assignments. While there remain many other factors that must be considered in selecting a person for a key role, having a corporate database of expertise ensures that all of the most technically qualified people can be considered.

9.3 Creating the company-wide database

In Chapters 6 and 7 a process was described for obtaining an accurate self-assessment of the current proficiency and development interests of each person in the company. If this data is then collected and entered into a common database, the tools are in place to address each of the needs described in the previous section.

In order for this tool to be effective in organizational planning and project staffing, three criteria must be met. First, a standard set of terminology must be adopted concerning the descriptions of each subject area and the expectations regarding proficiency. If a manager is not directly familiar with a person and that person's abilities, some method is needed to provide an objective understanding of what could or could not be expected of the person. The learning matrix and self-assessment provides this information, so here it is crucial that a common understanding has been applied across the corporation. If the subject descriptions and expectations versus proficiency level were well written they will fulfill this need. This was discussed previously in Chapters 3 and 4.

The second criterion for success follows directly from the first. The self-assessments must be accurate. As was previously discussed (Chapter 6) a clear, well-written description of expectations versus proficiency results in very accurate self-assessments.

The importance of the above two criteria may be summarized with reference to Figure 9.1. This is the Stress Analysis description previously given in Chapter 4. Assume that a manager is searching the company database, and identifies an engineer she or he has never met who is indicated as possessing Stress Analysis skills at Proficiency Level 3. The write-up shown in Figure 9.1 would give the manager a good understanding of the stress analysis capabilities possessed by this person.

The third criterion is typically the most difficult to address. As was discussed in Chapter 6, employees often perceive any kind of evaluation of their proficiency in punitive terms. This earlier discussion emphasized how placing the process in the employees' own hands reduces their discomfort and allows accurate personal assessments and individual development plans. However, as soon as the data is tabulated across a department or across the entire company, many employees will naturally become concerned. They now see themselves being directly compared, in

Stress Analysis

Proficiency Level 1

The engineer should be able to review the design and operation of a given component, identify the primary loads, load paths and necessary assumptions, and apply the fundamental equations of mechanics to obtain an estimate of design integrity. The engineer should also be able to determine an appropriate computational or experimental approach for obtaining more detailed analyses of the design.

An undergraduate engineering elective in stress analysis or mechanics of materials provides the required grounding. Guidance on specific problems is available through the company's Structural Analysis group.

Proficiency Level 2

The engineer should be able to critique a design to identify potential failure locations, produce calculations to estimate the likelihood of failure, and be able to propose design improvements and demonstrate their effectiveness through calculation. The engineer will be expected to be able to make appropriate assumptions, then select and utilize a model of sufficient but not excessive accuracy and complexity from available computational techniques. The engineer should also be able to identify simple experiments or measurements that can be used to test computational results or provide key information for improving the analysis.

This proficiency is achieved through demonstrated experience in application of stress analysis principles to a variety of design problems and should include at least basic skills in finite element analysis.

Proficiency Level 3

At this proficiency a key characteristic is experience and demonstrated success at applying the equations of mechanics in conjunction with simple experiments to solve structural problems. The engineer should also be able to apply the equations of mechanics to special cases such as elastic/plastic problems, three-dimensional load paths, and combinations of thermal and mechanical loading.

Several years of experience in the application of stress analysis principles to product design and development, over a wide range of components and problems, is required. Application of stress analysis should be supplemented by graduate course work.

Figure 9.1 Description and proficiency expectations for an engineering subject

writing, with their peers, and they will suspect that these comparisons will impact promotions and selections for the best positions (rightly so!). There is no simple way to address this problem. In implementing a company-wide database of employee proficiencies the first rule must be honesty and open communication. The employees should be clearly told exactly what the database is and how it is to be used. In reality, most employees should find this approach attractive, especially if they think about how the selection process is carried out in its absence. The judgments and selections are made either way and the database ensures that the selections are made based on objective data. But management must be sensitive to the fact that this is easier to describe than to buy into!

Once management is confident that the three criteria just described have been met, the individual assessments can be rolled up to create a company-wide database of employee capabilities and proficiencies. Depending on the number of employees and the amount of information that must be tracked, this database may be created manually, or automated with some form of relational database. The automated approach is preferred for most companies. Such software might be internally developed, or supplied by one of various human resources consulting firms that currently market such software.

Table 9.1 provides a simple example of how the resulting database might be constructed. Applying a numerical scale to the proficiency levels, as shown here, may be helpful in creating calculated measures of department or company strengths and weaknesses. In the example, a numerical average proficiency is calculated for each subject. The following paragraphs demonstrate the value of a department or company-wide database of engineering proficiencies with a variety of examples. The examples are based on a very small database (six measured subject areas and seven engineers); it should be apparent to the reader that these same techniques can be applied easily to a much larger database.

Turning first to a comparison of the department's strengths in Subjects 'B' and 'C' it is noted that exactly the same average proficiency is calculated for each of these subjects. However, in the case of Subject 'B' this average is based on a wide variety of proficiencies, with one person in the department showing a very high proficiency; in the case of Subject 'C' the entire department shows a modest proficiency. Based on this data one could draw the following conclusions:

Table 9.1 Example of a department proficiency assessment

	Subject A	Subject B	Subject C	Subject D	Subject E	Subject F
Employee 1	4	3	3	5	2	4
Employee 2	5	5	3	2	3	3
Employee 3	3	1	3	1	1	2
Employee 4	2	2	2	1	1	3
Employee 5	5	3	3	4	2	4
Employee 6	4	3	2	3	2	3
Employee 7	3	2	3	1	2	2
Average	3.71	2.71	2.71	2.57	1.85	3.00
High	5	5	3	5	3	4
Low	2	1	2	1	1	2

1 The department is currently capable of completing tasks requiring the highest proficiency in Subject 'B.' A mentor is available who could be used to guide others to greater proficiency.
2 The department is at risk of a significant drop in capability in subject area 'B' as only one person possesses the high proficiency – if she or he resigns, retires, or transfers, the capabilities of the department plummet. A succession plan is needed.
3 Almost everyone in the department possesses some capability in Subject 'C,' however no one has the capability to perform advanced level work in this area. A couple of questions must be asked to determine whether this situation needs to be changed. First, based on the work done by the department, do they need more advanced capability in this subject area, or are their applications of the subject so straightforward that the current proficiencies are all that is required? If more advanced proficiencies are needed, are they needed within the department, or is there another department that provides this capability as a supporting service?

The department shows a high degree of capability in Subject 'A.' This looks very attractive, but may also point to an area where someone is ready for a career move. Based on this subject alone, the department would not be at risk if any one of the engineers were to transfer to a new assignment. One would then have to look at additional subjects to determine who could transfer to another role while keeping any disruption of the work to a minimum. For example, looking at Subject 'A' alone one might suggest that Employee '2' or Employee '5' could bring their proficiency in this subject to another department. However, Employee '2' is currently the leading expert in Subject 'B,' and any transfer prior to implementing a succession plan would be fraught with problems. On the other hand, the transfer of Employee '5' would not cause such problems, as that person's other strongest proficiencies are in areas where the department possesses a great deal of expertise.

Based on this table it appears that Employees '2' and '3' are perhaps the newest members of the company. They should be seen as the technical contributors of the future and their individual development plans should reflect an emphasis on building proficiencies that best blend department needs with individual career interests.

Subject 'E' is one in which the department shows a serious weakness. One or two people should be selected for a development focus in this area, unless stronger proficiency in this area is not needed within the department (based on the nature of their work, or the availability of the proficiency in a supporting department).

9.4 Summary

- There are several ways in which an accurate company-wide understanding of proficiencies will be beneficial. Among them are the following:
  - providing a clear snapshot of department proficiencies;
  - succession planning for retirements, transfers, or resignations;
  - rapid and intelligent staffing of new projects;
  - ensuring that the best choices are made in internal staffing selections (employee transfers);
  - ensuring that staff selections fully consider employee diversity.
- The needed company-wide database of employee proficiencies is a natural extension of the learning model presented here. The data has been collected, and needs only to be rolled up into a single database.
- Three critical criteria must be addressed:
  - standardized terminology must be used throughout the corporation – everyone involved must have a clear, common understanding of what can be expected of an individual possessing a particular proficiency in a given subject;
  - the self-assessments must be accurate. It has previously been emphasized that if the proficiency expectations for a given subject have been well defined, an accurate self-assessment will directly follow;
  - the employees must not view proficiency tracking in punitive terms. This is an especially difficult challenge to meet, but must be ensured if the database is to be taken to the organizational level and used for staffing selections.
- This chapter concluded with several examples of how the department or corporate-wide database can be used to make decisions regarding employee development, project staffing and succession planning.

Chapter 10

Evaluating program effectiveness

10.1 Introduction

This brief chapter addresses an admittedly very different subject and the reader might question its inclusion in this book. The subject is that of measuring the effectiveness of training and development programs. There are two reasons for introducing the topic at this point. First, it is a topic that receives a great deal of attention in industry – in the continuing effort to control costs and remain competitive, corporate leaders are often found asking their training directors to demonstrate the cost-effectiveness of programs. This reason alone is not sufficient to include the subject in a book discussing tools for advanced learning models. But the second reason is that the tools that have been described in this book can easily be used to make a nice contribution to training evaluation.

The next section provides a brief review of training effectiveness measures – what it is that is being measured, and why. Sections 10.3 and 10.4 then describe a simple extension of the advanced learning tools to provide a measure of program effectiveness.

10.2 Review of Kirkpatrick's four-level model

As the curriculum of internal courses and professional development programs expands, it becomes more and more important to accurately evaluate the effectiveness of each of these programs. Kirkpatrick's four-level model is widely accepted in defining the various types of assessments that can be made, and will be adopted for this discussion. The four levels are briefly described as follows:

- Level I – Was the participant satisfied?
- Level II – Did the participant gain new knowledge or skills?

- Level III – Are the participant's new knowledge or skills resulting in changes in the work she or he is doing?
- Level IV – Are the work changes improving the cost-effectiveness of the organization?

Course evaluations given at the end of the course may provide an accurate Level I assessment, and it is generally believed that if the same (or similar) questions are asked several months later, they may provide some Level II information. Pre- and post-testing may provide Level II information, but has many pitfalls and is seldom used in industry. The use of Level IV evaluations (even if accurate calculations can be made) is widely debated in industrial organizations because the resulting conclusions depend on many factors beyond the training program itself. Level IV evaluation is generally considered to be possible for only the most closely defined training – typically that pertaining to psychomotor skills and resulting measurable improvements in productivity. Such evaluation is of little applicability to an engineering organization. The goal of most industrial concerns is therefore to develop accurate Level II or Level III training effectiveness evaluations.

10.3 Applying the available tools

The learning matrices and subject descriptions presented in this book allow a unique approach to achieving accurate Level II, and arguably Level III, evaluations without using testing.

Recall that on at least an annual basis each employee conducts a self-assessment on proficiencies relative to a set of engineering subjects. As a result of new learning the employee has experienced throughout the year, in successive years the proficiency identified for particular subjects will increase. Based on the self-evaluation, the employee then works with the supervisor to create an individual development plan for the coming year. While the initial basis of the process is self-evaluation, the results of the self-evaluation are then reviewed with the supervisor in assembling the next year's individual development plan. An upward adjustment in the quantified proficiency, and hence its removal from the development plan (or adjustment for still greater proficiency), must be agreed upon by the supervisor. Presumably such agreement is based on the supervisor's observation that the employee's ability to contribute to the organization has improved. Such a determination is exactly what is being made in a Level III evaluation. This looks promising, but how can the available tools (the learning matrix database and individual development plan) be used to provide a Level III program evaluation?

A very simple survey can be used in conjunction with the employees' self-assessment. As the self-assessments and development plans are completed, the employees are asked to identify the subject areas where their proficiencies have increased since their last review. For each subject that is identified in this exercise, employees are then asked to indicate their prior and current proficiency levels, and to briefly describe how the proficiency was improved.

The beauty of this approach lies in how little the employees are being asked to do. Their time faces many competing pressures, and neither the employees nor their supervisors react well to extensive forms or surveys that must be filled out (not to mention tests that might be taken!).

10.4 Summarizing the resulting data

Implementing the survey just described results in some very straightforward and quite useful information. The information consists simply of a listing of courses and development programs, and a raw number stating how many times each particular program was identified. These numbers must then be normalized based on the total number of people who participated in a given program during the year over which the survey is conducted. In other words, if a seminar was attended by 100 people over the previous twelve months, and it was identified in ten surveys, it should be given a normalized value of 0.1. This is necessary to ensure that the results are not skewed by the magnitude of a training program. The following example demonstrates the importance of this step.

A mandatory corporate training seminar on a new procedure was attended by 2000 people. In the same year, a world-renowned consultant was brought in to conduct a specialized, hands-on engineering seminar for ten people. At the end of the year all of the employees were asked to identify the programs that contributed to their improved proficiencies. Although the mandatory seminar was very poor, ten of the 2000 people who learned about the new procedure actually found it helpful. Eight of the ten people who attended the specialized seminar listed it on their surveys. The new corporate training director tallied up all the totals, and found the procedure seminar to be cited more frequently than any other. Delighted at the databased conclusion he was able to draw, he immediately started outlining next year's corporate procedure seminar.

While this example might be a bit extreme it demonstrates the need to ensure that programs are assessed relative to their overall size. While no one, after conducting such a survey, would question the fact that the specialized seminar was effective, the skewed results pertaining to the procedure seminar will result in great wastes of time, money and employee patience!

Having normalized the results based on program size, the resulting tabulation provides a ranking of both seminars and on-the-job development programs (work shadowing, one-on-one mentoring, etc.), based on the number of times each program was identified in the surveys. Any program receiving several citations can generally be considered as effectively contributing to employee development. It is those programs that are not cited that should be further scrutinized for their effectiveness. Admittedly this approach provides a rather open-ended identification of ineffective development programs – they are not directly being evaluated, but are being identified by their absence. The survey results provide a strong Level III affirmation of the programs that have been cited.

Finally, it should again be emphasized that this approach identifies not only formal training classes, but virtually any of the methods by which learning takes place. This is a very valuable result, and one that cannot easily be otherwise achieved. The company that implements this survey gains valuable information concerning how their employees learn – information that will be quite useful in creating the learning environment that further emphasizes the most effective approaches. This is especially useful as learning is focused on the higher proficiencies, where classroom approaches play a smaller role, and continued development relies more and more on the techniques that were described in Chapter 4. We have come full circle in now emphasizing evaluation techniques that address the advanced, experiential learning.

10.5 Summary

- Another potential benefit that can be realized with this experiential learning model is in the evaluation of training program effectiveness.
- Excellent data can be gathered by asking employees to respond to the following questions as they put together their individual development plans:
  - identify the subject areas where their proficiencies have increased since their last review;
  - for each of these subjects, describe how the proficiency was improved.
- This simple questionnaire will provide a listing of training programs as well as other mechanisms for learning (such as those described in Chapter 4) that have been identified as helpful in advancing proficiency.
- As this data is gathered from employees around the corporation, the number of times a particular program is cited must be normalized based on the number of people participating in the given program.
- The result of this process will be a listing of programs that have been identified as helpful in advancing employee proficiencies.
- Ineffective programs will be identified by their absence from this list.

Chapter 11

There is no such thing as a free lunch

The purpose of this book has been to provide a learning model that can be used in the workplace to

- place the focus on continued development of advanced proficiencies in subjects critical to the company's success;
- de-emphasize expensive corporate training efforts that focus on introductory proficiencies;
- place the responsibility for continued learning with each individual employee, while recognizing the important roles played by supervisors and upper management;
- provide guidance for, and emphasis on, learning as part of the job and not a separate activity;
- create a framework from which organizational assessment and succession planning can be accomplished based on sound data.

It is hoped that through the examples and discussion presented here, a convincing case has been made for the beauty and simplicity of this alternative model for learning in the workplace.

It must also be emphasized that this model requires investments of its own. Creating an accurate and useful set of learning matrices requires a considerable investment of time and resources, and the unwavering commitment of management. The learning matrices are the critical foundation upon which the model is implemented. Any shortcuts or carelessness in the creation of these matrices will quickly undermine the entire effort. Just think about how much of what we've discussed has been based on accurate employee proficiency assessments, and a common understanding of the definitions and proficiency expectations for each subject.

However, once the learning matrices have been completed, their maintenance in all but the most rapidly changing technology areas will be relatively low, and the gains seen by the organization will be tremendous

– rapid project staffing, excellent succession planning, a solid basis to measure training program effectiveness, are all direct benefits that can be achieved with this model. These gains are in addition to those resulting from employees seeing learning as part of what they do on the job each day.

References

BLOOM, B.S. (Ed.): 'Taxonomy of Educational Objectives: The Classification of Educational Goals, Handbook I: Cognitive Domain' (David McKay Company, Inc., New York, 1956)

CAFFARELLA, R.S., and BARNETT, B.G.: 'Characteristics of Adult Learners and Foundations of Experiential Learning,' in JACKSON, L. and CAFFARELLA, R. (Eds.): 'Experiential Learning: A New Approach' (Jossey-Bass Publishers, San Francisco, 1994) pp. 29–41

JACKSON, L., and MacISAAC, D: 'Introduction to a New Approach to Experiential Learning,' in JACKSON, L. and CAFFARELLA, R. (Eds.): 'Experiential Learning: A New Approach' (Jossey-Bass Publishers, San Francisco, 1994) pp. 17–28

KNOWLES, M.S.: 'The Modern Practice of Adult Education: From Pedagogy to Andragogy' (Cambridge Book Company, New York, 1980)

KNOX, A.B.: 'Helping Adults Learn' (Jossey-Bass Publishers, San Fransisco, 1986)

KOLB, D.A.: 'Experiential Learning: Experience as the Source of Learning and Development' (Prentice Hall, Englewood Cliffs, NJ, 1984)

LEE, P. and CAFFARELLA, R.S.: 'Methods and Techniques for Engaging Learners in Experiential Learning Activities,' in JACKSON, L. and CAFFARELLA, R. (Eds.): 'Experiential Learning: A New Approach' (Jossey-Bass Publishers, San Francisco, 1994) pp. 43–54

WEST, C.K., FARMER, J.A., and WOLFF, P.M.: 'Intructional Design: Implications from Cognitive Science' (Prentice Hall, Englewood Cliffs, NJ, 1991)

WRIGHT, M.D., and TILLMAN, T.: 'Career Competency Modeling of Multi-Industry Manufacturing Engineering Positions' Pacific Conference on *Manufacturing and Management*, 2000

Index

Note: For topics referenced on more than one page, bold type is used to indicate the primary references to the topic.